T0260499

The Normal and the Pathological

Translated by Carolyn R. Fawcett
in collaboration with Robert S. Cohen

The Normal and the Pathological

Georges Canguilhem

with an introduction by

Michel Foucault

ZONE BOOKS · NEW YORK

1991

ZONE BOOKS
633 Vanderbilt Street
Brooklyn, NY 11218

First Paperback Edition
Seventh Printing 2022

Originally published in 1966 as *Le Normal et le pathologique* © 1966 Presses Universitaires de France. Originally published in the English language and © 1978 by D. Reidel Publishing Company, Dordrecht, Holland.

Printed in the United States of America.

Distributed by Princeton University Press,
Princeton, New Jersey, and Woodstock, United Kingdom

Library of Congress Cataloging-in-Publication Data

Canguilhem, Georges, 1904–
The normal and the pathological.
Translation of: Le normal et le pathologique.
Reprint. Originally published: On the normal and the pathological. Dordrecht, Holland, Boston: D. Reidel, © 1978.
Includes bibliographies and index.
1. Pathology—Philosophy. 2. Medicine—Philosophy. I. Title.
RB112.C313 1989 616.07'01
ISBN 978-0-942299-59-5 (pbk.)

88-20626

Contents

Introduction

by

Michel Foucault

Everyone knows that in France there are few logicians but many historians of science; and that in the "philosophical establishment" – whether teaching or research oriented – they have occupied a considerable position. But do we know precisely the importance that, in the course of these past fifteen or twenty years, up to the very frontiers of the establishment, a "work" like that of Georges Canguilhem can have had for those very people who were separated from, or challenged, the establishment? Yes, I know, there have been noisier theaters: psychoanalysis, Marxism, linguistics, ethnology. But let us not forget this fact which depends, as you will, on the sociology of French intellectual environments, the functioning of our university institutions or our system of cultural values: in all the political or scientific discussions of these strange sixty years past, the role of the "philosophers" – I simply mean those who had received their university training in philosophy departments – has been important: perhaps too important for the liking of certain people. And, directly or indirectly, all or almost all these philosophers have had to "come to terms with" the teaching and books of Georges Canguilhem.

From this, a paradox: this man, whose work is austere, intentionally and carefully limited to a particular domain in the history of science, which in any case does not pass for a spectacular discipline, has somehow found himself present in discussions where he himself took care never to figure. But take away Canguilhem and you will no longer understand much about Althusser, Althusserism and a whole series of discussions which have taken place among French Marxists; you will no longer grasp what is specific to sociologists such as Bourdieu, Castel, Passerson and what marks them so strongly within sociology; you will miss an entire aspect of the theoretical work done by psychoanalysts, particularly by the followers of Lacan. Further, in the entire discussion of ideas which preceded or followed the movement of '68, it is easy to find the place of those who, from near or from afar, had been trained by Canguilhem.

Without ignoring the cleavages which, during these last years after the end of the war, were able to oppose Marxists and non-Marxists, Freudians and non-Freudians, specialists in a single discipline and philosophers, academics and non-academics, theorists and politicians, it does seem to me that one could find another dividing line which cuts through all these oppositions. It is the line that separates a philosophy of experience, of sense and of subject and a philosophy of knowledge, of rationality and of concept. On the one hand, one network is that of Sartre and Merleau-Ponty; and then another is that of Cavaillès, Bachelard and Canguilhem. In other words, we are dealing with two modalities according to which phenomenology was taken up in France, when quite late – around 1930 – it finally began to be, if not known, at least recognized. Contemporary philosophy in France began in those years. The lectures on transcendental phenomenology delivered in 1929 by Husserl (translated by Gabrielle Peiffer and Emmanuel Levinas as *Méditations cartésiennes*, Paris, Colin, 1931; and by Dorion Cairns as *Cartesian Meditations*, The Hague, Nijhoff, 1960) marked the moment: phenomenology entered France through that text. But it allowed of two readings: one, in the direction of a philosophy of the subject – and this was Sartre's article on the "Transcendance de L'Ego" (1935) and another which went back to the founding principles of Husserl's thought: those of formalism and intuitionism, those of the theory of science, and in 1938 Cavaillès's two theses on the *axiomatic method* and the *formation of set theory*. Whatever they may have been after shifts, ramifications, interactions, even rapprochements, these two forms of thought in France have constituted two philosophical directions which have remained profoundly heterogeneous.

On the surface the second of these has remained at once the most theoretical, the most bent on speculative tasks and also the most academic. And yet it was this form which played the most important role in the sixties, when a "crisis" began, a crisis concerning not only the University but also the status and role of knowledge. We must ask ourselves why such a mode of reflection, following its own logic, could turn out to be so profoundly tied to the present.

Undoubtedly one of the principal reasons stems from this: the history of science avails itself of one of the themes which was introduced almost surreptitiously into late eighteenth century philosophy: for the first time rational thought was put in question not only as to its nature, its foundation, its powers and its rights, but also as to its history and its geography; as to its immediate past and its present reality; as to its time and its place. This is the question which Mendelssohn and then Kant tried to answer in 1784 in the *Berlinische Monatschrift*: "Was ist Aufklärung?" (What is Enlightenment?). These two texts inaugurated a "philosophical journalism" which, along with university teaching, was one of the major forms of the institutional implantation of philosophy in the nineteenth century (and we know how fertile it sometimes was, as in the 1840s in Germany). They also opened philosophy up to a whole historico-critical dimension. And this work always involves two objectives which in fact, cannot be dissociated and which incessantly echo one another: on the one hand, to look for the moment (in its chronology, its constituent elements, its historical conditions) when the West first asserted the autonomy and sovereignty of its own rationality: the Lutheran Reformation, the "Copernican Revolution," Cartesian philosophy, the Galilean mathematization of nature, Newtonian physics. On the other hand, to analyze the "present" moment and, in terms of what was the history of this reason as well as of what can be its present balance, to look for that relation which must be established with this founding act: rediscovery, taking up a forgotten direction, completion or rupture, return to an earlier moment, etc.

Undoubtedly we should ask why this question of the Enlightenment, without ever disappearing, had such a different destiny in Germany, France and the Anglo-Saxon countries; why here and there it was invested in such different domains and according to such varied chronologies. Let us say in any case that German philosophy gave it substance above all in a historical and political reflection on society (with one privileged moment: the Reformation; and a central problem: religious experience in its relation with the economy and the state); from the Hegelians to the Frankfurt School and to Lukács, Feuerbach, Marx, Nietzsche and Max Weber it bears witness to this. In France it is the history of science which has above all served to support the philosophical question of the Enlightenment: after all, the positivism of Comte and his successors was one way of once again taking up the questioning by Mendelssohn and Kant on the scale of a general history of societies. Knowledge belief; the scientific form of knowledge and the religious contents of representation; or the transition from the pre-scientific or scientific; the constitution of a rational way of knowing on the basis of traditional experience; the appearance, in the midst of a history of ideas and beliefs, of a type of history suitable to scientific knowledge; the origin and threshold of rationality – it is under this form, through positivism (and those opposed to it), through Duhem, Poincaré, the noisy debates on scientism and the academic discussions about medieval science, that the question of the Enlightenment was brought into France. And if phenomenology, after quite a long period when it was kept at the border, finally penetrated in its turn, it was undoubtedly the day when Husserl, in the *Cartesian Meditations* and the *Crisis* (*The Crisis of European Sciences and Transcendental Phenomenology*, translated by David Carr, Evanston, Ill., Northwestern University Press, 1970), posed the question of the relations between the "Western" project of a universal development of reason, the positivity of the sciences and the radicality of philosophy.

If I have insisted on these points, it is to show that for a century and a half the history of science in France carried philosophical stakes within itself which are easily recognized. Works such as those of Koyré, Bachelard or Canguilhem could indeed have had as their centers of reference precise, "regional," chronologically well-defined domains in the history of science but they have functioned as important centers of philosophical elaboration to the extent that, under different facets, they set into play this question of the Enlightenment which is essential to contemporary philosophy.

If we were to look outside of France for something corresponding to the work of Cavaillès, Koyré, Bachelard and Canguilhem, it is undoubtedly in the Frankfurt School that we would find it. And yet, the styles are quite different: the ways of doing things, the domains treated. But in the end both pose the same kind of questions, even if here they are haunted by the memory of Descartes, there by the ghost of Luther. These questionings are those which must be addressed to a rationality which makes universal claims while developing in contingency; which asserts its unity and yet proceeds only by means of partial modification when not by general recastings; which authenticates itself through its own sovereignty but which in its history is perhaps not dissociated from inertias, weights which coerce it, subjugate it. In the history of science in France as in German critical theory, what we are to examine essentially is a reason whose autonomy of structures carries with itself the history of dogmatisms and despotisms – a reason which, consequently, has the effect of emancipation only on the condition that it succeeds in freeing itself of itself.

Several processes, marking the second half of the twentieth century, have led to the heart of contemporary preoccupations concerning the question of the Enlightenment. The first is the importance acquired by scientific and technical rationality in the development of the productive forces and the play of political decisions. The second is the very history of a "revolution" whose hope, since the close of the eighteenth century, had been borne by a rationalism to which we are entitled to ask, what part it could have in the effects of a despotism where that hope was lost.

The third and last is the movement by which, at the end of the colonial era, people began to ask the West what rights its culture, its science, its social organization and finally its rationality itself could have to laying claim to a universal validity: is it not a mirage tied to an economic domination and a political hegemony? Two centuries later the Enlightenment returns: but not at all as a way for the West to become conscious of its actual possibilities and freedoms to which it can have access, but as a way to question the limits and powers it has abused. Reason – the despotic enlightenment.

Let us not be surprised that the history of science, above all in the particular form given it by Georges Canguilhem, could have occupied so central a place in contemporary discussions in France, even if his role has remained rather hidden.

In the history of science, such as it was practiced in France, Georges Canguilhem brought about a significant shift. Broadly speaking, the history of science concerned itself by preference, if not exclusively, with disciplines which were "noble" in terms of the antiquity of their foundation, their high degree of formalization and their fitness for mathematization; in terms of the privileged position they occupied in the positivist hierarchy of the sciences. To remain close to these sciences which, from the Greeks to Leibniz, had, in short, been an integral part of philosophy, the history of science hid what it believed it was obliged to forget: that it was not philosophy. Canguilhem has focused almost all his work on the history of biology and medicine, knowing full well that the theoretical importance of the problems raised by the development of a science are not perforce in direct proportion to the degree of formalization reached by it. Thus he brought the history of science down from the heights (mathematics, astronomy, Galilean mechanics, Newtonian physics, relativity theory) toward the middle regions where knowledge is much less deductive, much more dependent on external processes (economic stimulations or institutional supports) and where it has remained tied much longer to the marvels of the imagination.

But in bringing about this shift, Canguilhem did more than assure the revaluation of a relatively neglected domain. He did not simply broaden the field of the history of science: he recast the discipline itself on a certain number of essential points:

1. He took up again the theme of "discontinuity" – an old theme which stood out very early, to the point of being contemporary,

or almost, with the birth of the history of science. What marks such a history, Fontenelle said, is the sudden formation of certain sciences "starting from nothing"; the extreme rapidity of some progress which was hardly expected; the distance separating scientific knowledge from "common usage" and the motives which could stimulate scientists; and furthermore, the potential form of this history which does not stop recounting the battles against "prejudices," "resistances" and "obstacles."[1] In taking up this same theme elaborated by Koyré and Bachelard, Canguilhem insists that for him marking discontinuities is neither a postulate nor a result, but rather a "way of doing," a process which is an integral part of the history of science because it is summoned by the very object which must be treated by it. In fact, this history of science is not a history of the true, of its slow epiphany; it would not be able to claim that it recounts the progressive discovery of a truth "inscribed forever in things or in the intellect," except to imagine that contemporary knowledge finally possesses it so completely and definitively that it can start from it to measure the past. And yet the history of science is not a pure and simple history of ideas and the conditions in which they appeared before being obliterated. In the history of science the truth cannot be given as acquired, but one can no longer economize on a relation to the truth and the true–false opposition. It is this reference to the "true–false" which gives this history its specificity and importance. In what form? By conceiving that one is dealing with the history of "truthful discourses," that is, discourses which rectify, correct themselves and which effect on themselves a whole work of elaboration finalized by the task of "speaking true." The historical tie which the different moments of science can have with one another necessarily has this form of discontinuity constituted by the alterings, reshapings, elucidations of new foundations, changes in scale, the transition to a new kind of object – "the perpetual revision of contents through thorough examination and amendment," as Cavaillès said. Error is not eliminated by the muffled force of a truth which gradually emerges from the shadow but by the formation of a new way of "speaking true."[2] One of the conditions of possibility because of which a history of science was formed at the beginning of the eighteenth century was, as Canguilhem notes, the awareness that there had been recent scientific "revolutions": that of algebraic geometry and the infinitesimal calculus, of Copernican and Newtonian cosmology.[3]

2. Whoever says "history of truthful discourse" also says recurrent method, not in the sense where the history of science would say: let the truth be finally recognized today, how long has one foreseen it, what paths had to be followed, what errors averted to discover it and prove it? But in the sense that the successive transformations of this truthful discourse continuously produce reshapings of their own history; what had for a long time remained a dead end, today becomes an exit; a "side" attempt becomes a central problem around which all the others gravitate; a slightly divergent step becomes a fundamental break: the discovery of non-cellular fermentation – a "side" phenomenon during the reign of Pasteur and his microbiology – marked an essential break only when the physiology of enzymes developed.[4] In short, the history of discontinuities is not acquired once and for all; it is itself "impermanent" and discontinuous.

Must we conclude from this that science spontaneously makes and remakes its own history at every instant, to the point that the only authorized historian of a science could be the scientist himself, reconstituting the past of what he was engaged in doing? The problem for Canguilhem is not a matter of a profession: it is a matter of point of view. The history of science cannot be content with bringing together what past scientists were able to believe or demonstrate; a history of plant physiology is not written by amassing

> everything that people called botanists, physicians, chemists, horticulturists, agronomists, economists could write down, touching on their conjectures, observations or experiences with regard to the relations between structure and function for objects which are sometimes called grass, sometimes plants, sometimes vegetables.[5]

But one does not make history of science either by refiltering the past through the set of statements or theories valid now, thus disclosing in what was "false" the true to come, and in what was true, the error made manifest later on. Here is one of the fundamental points of Canguilhem's method: the history of science can consist in what it has that is specific only by taking into account the epistemological point of view between the pure historian and the scientist himself. This point of view is that which causes a "hidden, ordered progression" to appear through different episodes of scientific knowledge: this means that the processes of elimination and selection of statements, theories, objects are made at each instant in terms of a certain norm; and this norm cannot be identified with a theoretical structure or an actual paradigm because today's scientific truth is itself only an episode of it – let us say provisional at most. It is not by depending on a "normal science" in T.S. Kuhn's sense that one can return to the past and validly trace its history: it is in rediscovering the "norm" process, the actual knowledge of which is only one moment of it, without one being able, save for prophesying, to predict the future. This history of science, says Canguilhem quoting Suzanne Bachelard, can construct its object only "in an ideal space-time." And this space-time is given to the history of science neither by the "realist" time accumulated by the historian's erudition nor by the idealized space authoritatively cut out by today's science, but by the point of view of epistemology. The latter is not the general theory of all science or of every possible scientific statement; it is the search for normativity within different scientific activities, such that they have effectively been brought into play. Hence we are dealing with an indispensable theoretical reflection which a history of science can form for itself in a way different from history in general; and conversely, the history of science opens up the area for analysis which is indispensable in order for epistemology to be something other than the simple reproduction of schemes within a science at a given moment.[6] In the method used by Canguilhem, the elaboration of "discontinuist" analyses and the elucidation of the history of science/epistemology relation go hand in hand.

3. Now, in placing the life sciences within this historico-epistemological perspective, Canguilhem brings to light a certain number of essential traits which single out the development of these sciences; and for their historians they pose specific problems. One had been able to believe around the time of Bichat that between a physiology studying the phenomena of life and a pathology dedicated to the analysis of diseases, one was finally about to disentangle what had remained confused for a long time in the mind of those who were studying the human body in order to "cure" it; and that having thus been freed from every immediate care of practice and every value judgment as to the good and evil functioning of the organism, one was finally going to be able to develop a pure and rigorous "science of life." But it proved impossible to make up a science of the living being without having taken into account, as essential to its object, the possibility of disease, death, monstrosity, anomaly, error (even if genetics gives this last word a meaning completely different from that intended by eighteenth-century physicians when

they spoke of an error of nature). You see, the living being involves self-regulation and self-preservation processes; with increasing subtlety we can know the physico-chemical mechanisms which assure them: they nonetheless mark a specificity which the life sciences must take into account, save for themselves omitting what properly constitutes their object and their own domain.

Hence a paradoxical fact in the life sciences: it is that if the "scientificization" process is done by bringing to light physical and chemical mechanisms, by the constitution of domains such as the chemistry of cells and molecules or such as biophysics, by the utilization of mathematical models, etc., it has on the other hand, been able to develop only insofar as the problem of the specificity of life and of the threshold it marks among all natural beings was continually thrown back as a challenge.[7] This does not mean that "vitalism," which has circulated so many images and perpetuated so many myths, is true. It does not mean that this idea, which has been so often rooted in less rigorous philosophies, must constitute the invincible philosophy of biologists. It simply means that it has had and undoubtedly still has an essential role as an "indicator" in the history of biology. And this in two respects: as a theoretical indicator of problems to be solved (that is, what, in general, constitutes the originality of life without, in any way, constituting an independent empire in nature); as a critical indicator of reductions to be avoided (that is, all those which tend to ignore the fact that the life sciences cannot do without a certain position of value indicating preservation, regulation, adaptation, reproduction, etc.). "A demand rather than a method, a morality more than a theory."[8]

Enlarging on the point, we could say that the constant problem in all Canguilhem's work, from the *Essai sur le normal et le pathologique* of 1943 to *Idéologie et rationalité dans l'histoire des sciences de la vie* (Ideology and Rationality in the History of the Life

Sciences) of 1977, has been the relation between science of life and vitalism: a problem which he tackled both in showing the irreducibility of the problem of disease as a problem essential to every science of life, and in studying what has constituted the speculative climate, the theoretical context of the life sciences.

4. What Canguilhem studies in a privileged way in the history of biology is the "formation of concepts." Most of the historical investigations he has conducted turn on this constitution: the concept of reflex, environment, monster and monstrosity, cell, internal secretion, regulation. There are several reasons for this. First of all, it is because the role of a strictly biological concept is to cut out from the ensemble of the phenomena "of life" those which allow one, without reducing, to analyze the processes proper to living beings (thus, among all the phenomena of resemblance, disappearance, mingling, recurrence proper to heredity, the concept of "hereditary trait" has brought about a similar "cutting out"): there is no object pertinent to biological science unless it has been "conceived." But, on the other hand, the concept does not constitute a limit which cannot be transcended by analysis: on the contrary, it must give access to a structure of intelligibility such that elementary analysis (that of chemistry or physics) allows one to show up the specific processes of the living being (this same concept of the hereditary trait led to a chemical analysis of the mechanisms of reproduction). Canguilhem insists that an idea becomes a biological concept at the moment the reductive effects, which are tied to an external analogy, become obliterated for the benefit of a specific analysis of the living being; the concept of "reflex" was not formed as a biological concept when Willis applied the image of a reflected light ray to an automatic movement; but it did happen the day Prochaska could write it down in the analysis of sensorimotor functions and their centralization in relation to

the brain.[9] Canguilhem would undoubtedly allow one to say that the moment which must be considered strategically decisive in a history of physics is that of the formalization and constitution of the theory; but the moment that counts in a history of the biological sciences is that of the constitution of the object and the formation of the concept.

The life sciences call for a certain manner of making their history. In a singular fashion they also pose the philosophical question of knowledge.

Life and death are never in themselves problems of physics, although in his work even the physicist risks his own life or that of others; for him these are questions of morals or politics, not of science. As A. Lwoff said, lethal or not, for the physicist a genetic mutation is neither more nor less than the substitution of one nucleic acid base for another. But it is in this very difference that the biologist recognizes the mark of his object; and an object of a type to which he himself belongs, since he lives and he manifests the nature of the living being, he exercises it, he develops it in an activity of knowledge which must be understood as a "general method for the direct or indirect resolution of tensions between man and the environment." The biologist must grasp what makes life a specific object of knowledge and thereby what makes it such that there are at the heart of living beings, because they are living beings, some beings susceptible to knowing, and, in the final analysis, to knowing life itself.

Phenomenology asked of "actual experience" the original meaning of every act of knowledge. But can we not, or must we not look for it in the living being himself?

Canguilhem, through the elucidation of knowledge concerning life and the concepts which articulate this knowledge, wants to rediscover which of them belongs to the *concept of life*. That is, the concept insofar as it is one of the modes of this information which every living being levies on his environment and by means of which, on the other hand, he structures his environment. That man lives in a conceptually architectured environment does not prove that he has been diverted from life by some oversight or that a historical drama has separated him from it; but only that he lives in a certain way, that he has a relationship with his environment such that he does not have a fixed point of view of it, that he can move on an undefined territory, that he must move about to receive information, that he must move things in relation to one another in order to make them useful. Forming concepts is one way of living, not of killing life; it is one way of living in complete mobility and not immobilizing life; it is showing, among these millions of living beings who inform their environment and are informed from it outwards, an innovation which will be judged trifling or substantial as you will: a very particular type of information.

Hence the importance Canguilhem accords the meeting, in the life sciences, of the old question of the normal and the pathological with the set of notions that biology, in the course of the last decades, has borrowed from information theory: code, messages, messengers, etc. From this point of view *Le normal et le pathologique*, written in part in 1943 and in part in the period 1963–66, constitutes without any doubt the most important and the most significant of Canguilhem's works. Here we see how the problem of the specificity of life recently found itself bent in one direction where we meet some of the problems believed to belong in their own right to the most developed forms of evolution.

At the heart of these problems is that of error. For at life's most basic level, the play of code and decoding leaves room for chance, which, before being disease, deficit or monstrosity, is something like perturbation in the information system, something like a "mistake." In the extreme, life is what is capable of error. And it is perhaps this given or rather this fundamental eventuality which must be called to account concerning the fact that the question of anomaly crosses all of biology, through and through. We must also call it to account for mutations and the evolutionary processes they induce. We must also call it to account for this singular mutation, this "hereditary error" which makes life result, with man, in a living being who is never completely at home, a living being dedicated to "error" and destined, in the end, to "error." And if we admit that the concept is the answer that life itself gives to this chance, it must be that error is at the root of what makes human thought and its history. The opposition of true and false, the values we attribute to both, the effects of power that different societies and different institutions link to this division – even all this is perhaps only the latest response to this possibility of error, which is intrinsic to life. If the history of science is discontinuous, that is, if it can be analyzed only as a series of "corrections," as a new distribution of true and false which never finally, once and for all, liberates the truth, it is because there, too, "error" constitutes not overlooking or delaying a truth but the dimension proper to the life of men and to the time of the species.

Nietzsche said that truth was the most profound lie. Canguilhem, who is at once close to and far from Nietzsche, would say perhaps that on the enormous calendar of life, it is the most recent error; he would say that the true–false division and the value accorded truth constitute the most singular way of living which could have been invented by a life which, from its furthermost origin, carried the eventuality of error within itself. Error for Canguilhem is the permanent chance around which the history of life and that of men develops. It is this notion of error which allows him to join what he knows about biology to the way he works its history without ever having wanted, as was done at the time of evolutionism, to deduce the latter from the former. It is this notion which allows him to mark the relation between life and the knowledge of life, and to follow, like a red thread, the presence of value and norm.

This historian of rationalities, himself a "rationalist," is a philosopher of error: I mean that it is in starting from error that he poses philosophical problems, I should say, *the* philosophical problem of truth and life. Here we touch on what is undoubtedly one of the fundamental events in the history of modern philosophy: if the great Cartesian break posed the question of the relations between truth and subject, the eighteenth century, as far as the relations of truth and life are concerned, introduced a series of questions of which the *Critique of Judgment* and the *Phenomenology of Spirit* were the first great formulations. And from then on it was one of the stakes of philosophical discussions: is it that knowledge of life must be considered as nothing more than one of the regions which depends on the general question of truth, subject and knowledge? Or is it that it obliges us to pose this question differently? Is it that the entire theory of the subject must not be reformulated, since knowledge, rather than opening itself up to the truth of the world, is rooted in the "errors" of life? We understand why Canguilhem's thought, his work as a historian and philosopher, could have so decisive an importance in France for all those who, starting from different points of view (whether theorists of Marxism, psychoanalysis or linguistics), have tried to rethink the question of the subject. Phenomenology could indeed introduce the body, sexuality, death, the perceived world into the field of analysis; the Cogito remained central; neither the rationality of science nor the specificity of the life sciences could compromise its founding role. It is to this philosophy of meaning, subject and the experienced thing that Canguilhem has opposed a philosophy of error, concept and the living being.

Foreword

The present work unites two studies – one unpublished – on the same subject. It is first a re-edition of my doctoral thesis in medicine, made possible by the gracious consent of the Publications Committee of the Faculty of Letters at Strasbourg for this project of the Presses Universitaires de France. To those who conceived the project as well as to those who furthered its realization, I express here my heartfelt gratitude.

It is not for me to say whether this re-edition is necessary or not. It is true that my thesis was fortunate enough to arouse interest in medical as well as philosophical circles. I am left with the hope that it will not be judged now as being too out of date.

In adding some unpublished considerations to my first Essay (Section 1), I am only trying to furnish evidence of my efforts – if not my success – to preserve a problem, which I consider fundamental, in the same state of freshness as its everchanging factual data.

G.C.
1966

This revised edition contains corrections of some details and some supplementary footnotes indicated by an asterisk.

G.C.
1972

Section One

ESSAY ON SOME PROBLEMS CONCERNING THE NORMAL AND THE PATHOLOGICAL

(1943)

Preface to the Second Edition (1950)

This second edition of my doctoral thesis in medicine exactly reproduces the text of the first, published in 1943. This is by no means because of my own definitive satisfaction with it. But, on the one hand, the Publications Committee of the Faculty of Letters at Strasbourg – whom I very cordially thank for having decided to reprint my work – could not afford the expense involved in changing the text. On the other hand, the corrections or additions to this first essay will be found in a future, more general work. I would only like to indicate here those new readings, those criticisms which have been made, those personal reflections with which I could and should have enriched the first version of my essay.

To begin with, even in 1943 I could have pointed out what help I could find for the central theme of my exposition in works such as Pradines's *Traité de psychologie générale* and Merleau-Ponty's *Structure du comportement*. I could only indicate the second, discovered when my manuscript was in press. I had not yet read the first. Suffice it to recall the conditions for distributing books in 1943 in order to understand the difficulties of documentation at that time. Furthermore, I must confess I am not too sorry about them as I much prefer a convergence whose fortuitous character better emphasizes the value of intellectual necessity to an acquiescence, even fully sincere, in the view of others.

If I were to write this essay today I would have to devote a great deal of space to Selye's works and his theory of the state of organic alarm. This exposition could serve to mediate between Leriche's and Goldstein's theses (at first glance very different) of which I have the highest opinion. Selye established that the failures or irregularities of behavior as well as the emotions and fatigue they generate, produce through their frequent repetition, a structural modification of the adrenal cortex analogous to that determined by the introduction of hormonal substances (whether impure or pure but in large doses) or toxic substances into the internal environment. Every organic state of disordered tension, all behavior of alarm and stress, provoke adrenal reaction. This reaction is "normal" with regard to the action and effects of corticosterone in the organism. Moreover, these structural reactions, which Selye calls adaptation reactions and alarm reactions, involve the thyroid or hypophysis as well as the adrenal gland. But these normal (that is biologically favorable) reactions end up wearing out the organism in the case of abnormal (that is statistically frequent) repetitions of situations which generate the alarm reaction. In certain individuals, then, disadaptation diseases are set up. The repeated discharge of corticosterone provokes either functional disturbances such as vascular spasm and hypertension or morphological lesions such as stomach ulcer. Hence in the populations of English villages subjected to air raids in the last war, a notable increase in cases of gastric ulcer was observed.

If these facts are interpreted from Goldstein's point of view, disease will be seen in catastrophic behavior; if they are interpreted from Leriche's point of view, disease will be seen in the determination of histological anomaly by physiological disorder. These two points of view are not mutually exclusive, far from it.

Likewise, in the case of my references to the problems of

teratogenesis, today I would draw a great deal on Etienne Wolff's works on *Les changements de sexe* and *La science des monstres*. I would insist more on the possibility and even the obligation of enhancing the knowledge of normal formations by using knowledge about monstrous formations. I would propose more forcefully that there is not in itself an *a priori* ontological difference between a successful living form and an unsuccessful form. Moreover, can we speak of unsuccessful living forms? What lack can be disclosed in a living form as long as the nature of its obligations as a living being has not been determined?

I also should have taken into account – more than the approvals or confirmations which reached me from physicians, psychologists, such as my friend Lagache, professor at the Sorbonne, or biologists such as Sabiani and Kehl at the Algiers Faculty of Medicine – the criticism, at once comprehensive and firm, of Louis Bounoure of the Faculty of Sciences at Strasbourg. In his *L'autonomie de l'être vivant* Bounoure reproaches me with as much spirit as cordiality for yielding to the "evolutionist obsession" and considers, if I may say, with great perspicacity, the idea of the living being's normativity as a projection onto all of living nature of the human tendency toward transcendence. Whether it is legitimate or not to introduce History into Life (I am thinking here of Hegel and the problems raised by the interpretation of Hegelianism) is indeed a serious problem, at once biological and philosophical. Understandably this question cannot be tackled in a preface. At the least I want to say that it has not escaped my attention, that I hope to tackle it later, and that I am grateful to Bounoure for helping me to pose it.

Finally, it is certain that in expounding Claude Bernard's ideas today, I could not help taking into account the publication in 1947 by Dr. Delhoume of the *Principes de médecine expérimentale*, where Bernard is more precise than elsewhere in examining the problem of the individual relativity of the pathological fact. But essentially

I do not think that my judgment of Bernard's ideas would be modified.

In concluding I want to add that certain readers were surprised at the brevity of my conclusions and at the fact that they leave the philosophical door open. I must say that this was intentional. I had wanted to lay the groundwork for a future thesis in philosophy. I was aware of having sacrificed enough, if not too much, to the philosophical demon in a thesis in medicine. And so I deliberately gave my conclusions the appearance of propositions which were simply and moderately methodological.

Introduction

The problem of pathological structures and behaviors in man is enormous. A congenital clubfoot, a sexual inversion, a diabetic, a schizophrenic, pose innumerable questions which, in the end, refer to the whole of anatomical, embryological, physiological and psychological research. It is nevertheless our opinion that this problem must not be broken up and that the chances for clarifying it are greater if it is considered *en bloc* than if it is broken down into questions of detail. But for the moment we are in no position to maintain this opinion by presenting a sufficiently documented synthesis, which we do hope to work out one day. However, this publication of some of our research expresses not only this present impossibility but also the intention to mark successive phases in the inquiry.

Philosophy is a reflection for which all unknown material is good, and we would gladly say, for which all good material must be unknown. Having taken up medical studies some years after the end of our philosophical studies, and parallel to teaching philosophy, we owe some explanation of our intentions. It is not necessarily in order to be better acquainted with mental illnesses that a professor of philosophy can become interested in medicine. Nor is it necessarily in order to exercise a scientific discipline. We expected medicine to provide precisely an introduction to concrete human problems. Medicine seemed to us and still seems to us like a technique or art at the crossroads of several sciences, rather than, strictly speaking, like one science. It seemed to us that the two problems which concerned us, that of the relations between science and technology, and that of norms and the normal, had to profit from a direct medical education for their precise position and clarification. In applying to medicine a spirit which we would like to be able to call "unprejudiced," it seemed to us that, despite so many laudable efforts to introduce methods of scientific rationalization, the essential lay in the clinic and therapeutics, that is, in a technique of establishing or restoring the normal which cannot be reduced entirely and simply to a single form of knowledge.

The present work is thus an effort to integrate some of the methods and attainments of medicine into philosophical speculation. It is necessary to state that it is not a question of teaching a lesson, or of bringing a normative judgment to bear upon medical activity. We are not so presumptuous as to pretend to renovate medicine by incorporating a metaphysics into it. If medicine is to be renovated, it is up to physicians to do so at their risk and to their credit. But we want to contribute to the renewal of certain methodological concepts by adjusting their comprehension through contact with medical information. May no one expect more from us than we wanted to give. Medicine is very often prey and victim to certain pseudo-philosophical literature, not always unknown, it must be said, to doctors, in which medicine and philosophy rarely come out well. It is not our intention to bring grist to the mill. Nor do we intend to behave as an historian of medicine. If we have placed a problem in historical perspective in the first part of our book, it is only for reasons of greater intelligibility. We claim no erudition in biography.

A word on the boundaries of our subject. From the medical

point of view, the general problem of the normal and the pathological can be defined as a teratological problem and a nosological problem and this last, in its turn, as a problem of somatic nosology or pathological physiology, and as a problem of psychic nosology or pathological psychology. In the present exposition we want to limit ourselves very strictly to the problem of somatic nosology or pathological physiology, without, however, refraining from borrowing from teratology or pathological psychology this datum, that notion or solution, which would seem to us particularly suited to clarify the investigation or confirm some result.

We have also tried to set forth our conceptions in connection with the critical examination of a thesis, generally adopted in the nineteenth century, concerning the relations between the normal and the pathological. This is the thesis according to which pathological phenomena are identical to corresponding normal phenomena save for quantitative variations. With this procedure we are yielding to a demand of philosophical thought to reopen rather than close problems. Léon Brunschvicg said of philosophy that it is the science of solved problems. We are making this simple and profound definition our own.

Part One

Is the Pathological State Merely a Quantitative Modification of the Normal State?

CHAPTER I

Introduction to the Problem

To act, it is necessary at least to localize. For example, how do we take action against an earthquake or hurricane? The impetus behind every ontological theory of disease undoubtedly derives from therapeutic need. When we see in every sick man someone whose being has been augmented or diminished, we are somewhat reassured, for what a man has lost can be restored to him, and what has entered him can also leave. We can hope to conquer disease even if it is the result of a spell, or magic, or possession; we have only to remember that disease happens to man in order not to lose all hope. Magic brings to drugs and incantation rites innumerable resources for generating a profoundly intense desire for cure. Sigerist has noted that Egyptian medicine probably universalized the Eastern experience of parasitic diseases by combining it with the idea of disease-possession: throwing up worms means being restored to health [107, *120*].[1] Disease enters and leaves man as through a door.

A vulgar hierarchy of diseases still exists today, based on the extent to which symptoms can – or cannot – be readily localized, hence Parkinson's disease is more of a disease than thoracic shingles, which is, in turn, more so than boils. Without wishing to detract from the grandeur of Pasteur's tenets, we can say without hesitation that the germ theory of contagious disease has certainly owed much of its success to the fact that it embodies an ontological representation of sickness. After all, a germ can be seen, even if this requires the complicated mediation of a microscope, stains and cultures, while we would never be able to see a miasma or an influence. To see an entity is already to foresee an action. No one will object to the optimistic character of the theories of infection insofar as their therapeutic application is concerned. But the discovery of toxins and the recognition of the specific and individual pathogenic role of *terrains* have destroyed the beautiful simplicity of a doctrine whose scientific veneer for a long time hid the persistence of a reaction to disease as old as man himself. [For *terrain*, see glossary – Tr.]

If we feel the need to reassure ourselves, it is because one anguish constantly haunts our thoughts; if we delegate the task of restoring the diseased organism to the desired norm to technical means, either magical or matter of fact [*positive*] it is because we expect nothing good from nature itself.

By contrast, Greek medicine, in the Hippocratic writings and practices, offers a conception of disease which is no longer ontological, but dynamic, no longer localizationist, but totalizing. Nature (*physis*), within man as well as without, is harmony and equilibrium. The disturbance of this harmony, of this equilibrium, is called disease. In this case, disease is not somewhere in man, it is everywhere in him; it is the whole man. External circumstances are the occasion but not the causes. Man's equilibrium consists of four humors, whose fluidity is perfectly suited to sustain variations and oscillations and whose qualities are paired by opposites (hot/cold, wet/dry); the disturbance of these humors causes disease. But disease is not simply disequilibrium or discordance; it is, and perhaps most important, an effort on the part of nature to effect a new equilibrium in man. Disease is a generalized reaction designed to bring about a cure; the organism develops a disease in order to get well. Therapy must first tolerate and if necessary, reinforce these hedonic and spontaneously therapeutic reactions. Medical technique imitates natural medicinal action (*vis medicatrix naturae*). To imitate is not merely to copy an appearance: but to mimic a tendency and to extend an intimate movement. Of course, such a conception is also optimistic, but here the optimism concerns the way of nature and not the effect of human technique.

Medical thought has never stopped alternating between these two representations of disease, between these two kinds of optimism, always finding some good reason for one or the other attitude in a newly explained pathogenesis. Deficiency diseases and all infectious or parasitic diseases favor the ontological theory, while endocrine disturbances and all diseases beginning with *dys-* support the dynamic or functional theory. However, these two conceptions do have one point in common: in disease, or better, in the experience of being sick, both envision a polemical situation: either a battle between the organism and a foreign substance, or an internal struggle between opposing forces. Disease differs from a state of health, the pathological from the normal, as one quality differs from another, either by the presence or absence of a definite principle, or by an alteration of the total organism. This heterogeneity of normal and pathological states persists today in the naturalist conception, which expects little from human efforts to restore the norm, and in which nature will find the ways toward cure. But it proved difficult to maintain the qualitative modification separating the normal from the pathological in a conception which allows, indeed expects, man to be able to compel nature and bend it to his normative desires. Wasn't it said repeatedly after Bacon's time that one governs nature only by obeying it? To govern disease means to become acquainted with its relations with the normal state, which the living man – loving life – wants to regain. Hence the theoretical need, but a past due technique, to establish a scientific pathology by linking it to physiology. Thomas Sydenham (1624–1689) thought that in order to help a sick man, his sickness had to be delimited and determined. There are disease species just as there are animal or plant species. According to Sydenham there is an order among diseases similar to the regularity Isidore Geoffroy Saint-Hilaire found among anomalies. Pinel justified all these attempts at classification of disease [nosology] by perfecting the genre in his *Nosographie philosophique* (1797), which Daremberg described as more the work of a naturalist than a clinician [29, *1201*].

Meanwhile, Morgagni's (1682–1771) creation of a system of pathological anatomy made it possible to link the lesions of certain organs to groups of stable symptoms, such that nosographical classification found a substratum in anatomical analysis. But just as the followers of Harvey and Haller "breathed life" into anatomy by turning it into physiology, so pathology became a natural extension of physiology. (Sigerist provides a masterful summary of this evolution of medical ideas: see 107, *117–142*.) The end result of this evolutionary process is the formation of a theory of the relations between the normal and the pathological, according to which the pathological phenomena found in living organisms are nothing more than quantitative variations, greater or lesser according to corresponding physiological phenomena. Semantically, the pathological is designated as departing from the normal not so much by *a-* or *dys-* as by *hyper-* or *hypo-*. While retaining the ontological theory's soothing confidence in the possibility of technical conquest of disease, this approach is far from considering health and sickness as qualitatively opposed, or as forces joined in battle. The need to reestablish continuity in order to gain more knowledge for more effective action is such that the concept of disease would finally vanish. The conviction that one can scientifically restore the norm is such that in the end it annuls the pathological. Disease is no longer the object of anguish for the healthy man; it has become instead the object of study for the theorist of health. It is in pathology, writ large, that we can unravel the teachings of health, rather as Plato sought in the institutions of the State the larger and more easily readable equivalent of the virtues and vices of the individual soul.

In the course of the nineteenth century, the real identity of normal and pathological vital phenomena, apparently so different, and given opposing values by human experience, became a kind of scientifically guaranteed dogma, whose extension into the realms of philosophy and psychology appeared to be dictated by the authority biologists and physicians accorded to it. This dogma was expounded in France by Auguste Comte and Claude Bernard, each working under very different circumstances and with very different intentions. In Comte's doctrine the dogma is based on an idea taken (with explicit and respectful thanks) from Broussais. In Claude Bernard it is the conclusion drawn from an entire lifetime of biological experimentation, the practice of which is methodically codified in the famous *Introduction à l'étude de la médecine expérimentale*. In Comte's thought interest moves from the pathological to the normal, with a view to determining speculatively the laws of the normal; for it is as a substitute for biological experimentation – often impracticable, particularly on man – that disease seems worthy of systematic study. The identity of the normal and the pathological is asserted as a gain in knowledge of the normal. Bernard's interest moves from the normal to the pathological with a view toward rational action directed at the pathological; for it is as the foundation of an emphatically non-empirical therapeutics that knowledge of disease is sought by means of physiology and deriving from it. The identity of the normal and the pathological is asserted as a gain in remedying the pathological. Finally, in Comte the assertion of identity remains purely conceptual, while Claude Bernard tries to make this identity precise in a quantitative, numerical interpretation.

In calling such a theory a dogma we do not mean at all to disparage it, but rather to stress its scope and repercussions. Nor is it at all by chance that we decided to look to Comte and Bernard for the texts that determined its meaning. The influence of these two writers on nineteenth-century philosophy and science, and perhaps even more on literature, is considerable. It is well established that physicians are more willing to look for the philosophy of their art in literature than in medicine or philosophy themselves. Reading Littré, Renan and Taine has certainly inspired more medical careers than reading Richerand or Trousseau: it is a fact to be reckoned with that people generally enter medicine completely ignorant of medical theories, but not without preconceived notions about many medical concepts. The dissemination of Comte's ideas in medical, scientific and literary circles was the work of Littré and Charles Robin, first incumbent of the chair of histology at the Faculty of Medicine in Paris.[2] Their influence is felt most of all in the field of psychology. From Renan we learn:

> In studying the psychology of the individual, sleep, madness, delirium, somnambulism, hallucination offer a far more favorable field of experience than the normal state. Phenomena, which in the normal state are almost effaced because of their tenuousness, appear more palpable in extraordinary crises because they are exaggerated. The physicist does not study galvanism in the weak quantities found in nature, but increases it, by means of experimentation, in order to study it more easily, although the laws studied in that exaggerated state are identical to those

> of the natural state. Similarly human psychology will have to be constructed by studying the madness of mankind, the dreams and hallucinations to be found on every page of the history of the human spirit [99, *184*].

L. Dugas, in his study of Ribot, clearly showed the relationship between Ribot's methodological views and the ideas of Comte and Renan, his friend and protector [37, *21* and *68*]:

> Physiology and pathology, both physical and psychological, do not stand in contrast to each other as two opposites, but rather as two parts of the same whole.... The pathological method tends simultaneously toward pure observation and experimentation. It is a powerful means of investigation which has been rich in results. Disease is, in effect, an experiment of the most subtle order, instituted by nature itself in very precise circumstances by means unavailable to human skill: nature reaches the inaccessible [100].

Claude Bernard's influence on physicians between 1870 and 1914 is equally broad and deep, both directly through physiology and indirectly through literature, as established by the works of Lamy and Donald-King on the relations between literary naturalism and nineteenth-century biological and medical doctrines [68 and 34]. Nietzsche borrowed from Claude Bernard precisely the idea that the pathological is homogeneous with the normal. Quoting a long passage on health and sickness taken from *Leçons sur la chaleur animale* (Lectures on Animal Heat),[3] Nietzsche precedes it with the following statement:

> It is the value of all morbid states that they show us under a magnifying glass certain states that are normal – but not easily visible when normal.

These summary indications must suffice to show that the thesis whose meaning and importance we are trying to define has not been invented for the sake of the cause. The history of ideas cannot be superimposed perforce on the history of science. But as scientists lead their lives as men in an environment and social setting that is not exclusively scientific, the history of science cannot neglect the history of ideas. In following a thesis to its logical conclusion, it could be said that the modifications it undergoes in its cultural milieu can reveal its essential meaning.

We chose to center our exposition around Comte and Claude Bernard because these writers really played the role, half voluntarily, of standardbearer; hence the preference given them over so many others, who are cited to an equal extent and who could have been more vividly explained from one or another point of view.[4] For precisely the opposite reason, we decided to add the exposition of Leriche's ideas to that of Comte's and Bernard's. Leriche is discussed as much in medicine as in physiology – not the least of his merits. But it is possible that an examination of his ideas from an historical perspective will reveal unsuspected depth and significance. Without succumbing to a cult of authority, we cannot deny an eminent practitioner a competence in pathology excelling that of Comte and Claude Bernard. Moreover, as far as the problems examined here are concerned, it is not without interest that Leriche presently occupies the chair of medicine at the Collège de France made famous by Claude Bernard himself. Thus, the differences between them are only the more meaningful and valuable.

Chapter II

Auguste Comte and "Broussais's Principle"

Auguste Comte asserted the real identity of pathological phenomena and the corresponding physiological phenomena at three principal stages of his intellectual development: first, in the period leading up to the *Cours de philosophie positive*, characterized, at the beginning, by his friendship with Saint-Simon, with whom he severed relations in 1824;[5] second, the actual period of the positive philosophy; and third, the period of the *Systeme de politique positive*, which, in certain respects, is very different from the preceding one. Comte gave what he called Broussais's principle universal significance in the order of biological, psychological and sociological phenomena.

It was in 1828 that Comte took notice of Broussais's treatise *De l'irritation et de la folie* [On Irritation and Madness] and adopted the principle for his own use. Comte credits Broussais, rather than Bichat, and before him, Pinel, with having declared that all diseases acknowledged as such are only symptoms and that disturbances of vital functions could not take place without lesions in organs, or rather, tissues. But above all, adds Comte, "never before had anyone conceived the fundamental relation between pathology and physiology in so direct and satisfying a manner." Broussais described all diseases as consisting essentially "in the excess or lack of excitation in the various tissues above or below the degree established as the norm." Thus, diseases are merely the effects of simple changes in intensity in the action of the stimulants which are indispensable for maintaining health.

From then on Comte raised Broussais's nosological conception to the level of a general axiom. It would not be exaggerating to say that he accorded it the same dogmatic value as Newton's law or d'Alembert's principle. Certainly when he tried to link his fundamental sociological principle, "progress is nothing but the development of order," to some other more general principle which could verify it, Comte hesitated between Broussais's authority and d'Alembert's. He refers sometimes to d'Alembert's reduction of the laws of the propagation of movement to the laws of equilibrium [28, *I, 490–94*], sometimes to Broussais's aphorism. The positive theory of the changeability of phenomena

> is completely reduced to this universal principle and results from the systematic application of Broussais's great aphorism: every modification – whether natural or artificial – of the real order concerns only the intensity of the corresponding phenomena . . . ; despite variations in degree, phenomena always retain the same arrangement; every change in the actual *nature*, that is, class, of an object is recognized moreover as being contradictory [28, *III*, 71].

Little by little Comte practically claimed the intellectual paternity of this principle for himself by virtue of the fact that he applied it systematically, exactly as he at first thought that Broussais, having borrowed the principle from Brown, was able to claim it for himself because of the personal use he had made of it [28, *IV, App. 223*]. Here we must quote a rather long passage which would be weakened if summarized:

In the case of living beings, the judicious observation of disease forms a series of indirect experiments which is much more suitable than most direct experiments to throw light on explaining dynamic and even statistical notions. My philosophical Treatise did much to commend the nature and scope of such a procedure which leads to truly important gains in biology. It rests on the great principle, whose discovery I attribute to Broussais as it derives from the sum total of his works, although I alone constructed the general and direct formula. Until Broussais, the pathological state obeyed laws completely different from those governing the normal state, so that the exploration of one could have no effect on the other. Broussais established that the phenomena of disease coincided essentially with those of health from which they differed only in terms of intensity. This brilliant principle has become the basis of pathology, thus subordinated to the whole of biology. Applied in the opposite sense it explains and improves the great capacity of pathological analysis for throwing light on biological speculations. . . . The insights already gained from it can only give a faint idea of its ultimate efficacy. Those engaged in the encyclopedic task of compiling and classifying knowledge will extend Broussais's principle primarily to moral and intellectual activities where it has not yet received a worthy application, hence their diseases surprise or move us without instructing us. . . . In the general system of positive education, besides its direct usefulness for biological problems, this principle will be an appropriate logical preparation for analogous procedures in any science. The collective organism, because of its greater degree of complexity, has problems more serious, varied, and frequent than those of the individual organism. I do not hesitate to state that Broussais's principle must be extended to this point and I have often applied it to confirm or perfect sociological

> laws. But the analysis of revolutions could not illuminate the positive study of society without the logical initiation resulting, in this respect, from the simplest cases presented by biology [28, *I, 651–53*].

Here then is a principle of nosology vested with a universal authority that embraces the political order. Moreover, it goes without saying that it is this last projected application which confers the principle with all the value of which it is already capable, according to Comte, in the biological order.

The fortieth lecture of the *Cours de philosophie positive* – philosophical reflections on the whole of biology – contains Comte's most complete text on the problem now before us. It is concerned with showing the difficulties inherent in the simple extension of experimental methods, which have proved their usefulness in the physicochemical sphere, to the particular characteristics of the living:

> Any experiment whatever is always designed to uncover the laws by which each determining or modifying influence of a phenomenon effects its performance, and it generally consists in introducing a clear-cut change into each designated condition in order to measure directly the corresponding variation of the phenomenon itself [27, *169*].

Now, in biology the variation imposed on one or several of a phenomenon's conditions of existence cannot be random but must be contained within certain limits compatible with the phenomenon's existence. Furthermore, the fact of functional *consensus* proper to the organism precludes monitoring the relation, which links a determined disturbance to its supposedly exclusive effects, with sufficient analytical precision. But, thinks Comte, if we readily admit that the essence of experimentation lies not in the researcher's artificial intervention in the system of a phenomenon which he intentionally tends to disturb, but rather in the comparison between a control phenomenon and one altered with respect to any one of its conditions of existence, it follows that diseases must be able to function for the scientists as spontaneous experiments which allow a comparison to be made between an organism's various abnormal states and its normal state.

> According to the eminently philosophical principle which will serve from now on as a direct, general basis for positive pathology and whose definitive establishment we owe to the bold and persevering genius of our famous fellow citizen, Broussais, the pathological state is not at all radically different from the physiological state, with regard to which – no matter how one looks at it – it can only constitute a simple extension going more or less beyond the higher or lower limits of variation proper to each phenomenon of the normal organism, without ever being able to produce really new phenomena which would have to a certain degree any purely physiological analogues [27, *175*].

Consequently, every conception of pathology must be based on prior knowledge of the corresponding normal state, but conversely, the scientific study of pathological cases becomes an indispensable phase in the overall search for the laws of the normal state. The observation of pathological cases offers numerous, genuine advantages for actual experimental investigation. The transition from the normal to the abnormal is slower and more natural in the case of illness, and the return to normal, when it takes place, spontaneously furnishes a verifying counterproof. In addition, as far as man is concerned, pathological investigation is more fruitful than the necessarily limited experimental exploration. The scientific study of morbid states is essentially valid for all organisms, even plant life, and is particularly suited to the most complex and, therefore, the most delicate and fragile phenomena which direct experimentation, being too brusque a disturbance, would tend to distort. Here Comte was thinking of vital phenomena related to the higher animals and man, of the nervous and psychic functions. Finally, the study of anomalies and monstrosities conceived as both older and less curable illnesses than the functional disturbances of various plant or neuromotor apparatuses completes the study of diseases: the "teratological approach" [study of monsters] is added to the "pathological approach" in biological investigation [27, *179*].

It is appropriate to note, first, the particularly abstract quality of this thesis and the absence throughout of any precise example of a medical nature to suitably illustrate his literal exposition. Since we cannot relate these general propositions to any example, we do not know from what vantage point Comte states that the pathological phenomenon always has its analogue in a physiological phenomenon, and that it is nothing radically new. How is a sclerotic artery analogous to a normal one, or an asystolic heart identical to that of an athlete at the height of his powers? Undoubtedly, we are meant to understand that the laws of vital phenomena are the same for both disease and health. But then why not say so and give examples? And even then, does this not imply that analogous effects are determined in health and disease by analogous mechanisms? We should think about this example given by Sigerist:

> During digestion the number of white blood cells increases. The same is true at the onset of infection. Consequently this phenomenon is sometimes physiological, sometimes pathological, depending on what causes it [107, *109*].

Secondly, it should be pointed out that despite the reciprocal nature of the clarification achieved through the comparison of the normal with the pathological and the assimilation of the pathological and the normal, Comte insists repeatedly on the necessity of determining the normal and its true limits of variation first, before methodically investigating pathological cases. Strictly speaking, knowledge of normal phenomena, based solely on observation, is both possible and necessary without knowledge of disease, particularly based on experimentation. But we are presented with a serious gap in that Comte provides no criterion which would allow us to know what a normal phenomenon is. We are left to conclude that on this point he is referring to the usual corresponding concept, given the fact that he uses the notions of normal state, physiological state and natural state interchangeably [27, *175, 176*]. Better still, when it comes to defining the limits of pathological or experimental disturbances compatible with the existence of organisms, Comte identifies these limits with those of a "harmony of distinct influences, those exterior as well as interior" [27, *169*] – with the result that the concept of the normal or physiological, finally clarified by this concept of *harmony* amounts to a qualitative and polyvalent concept, still more aesthetic and moral than scientific.

As far as the assertion of identity of the normal phenomenon and the corresponding pathological phenomenon is concerned, it is equally clear that Comte's intention is to deny the qualitative difference between these two admitted by the vitalists. Logically, to deny a qualitative difference must lead to asserting a homogeneity capable of expression in quantitative terms. Comte is undoubtedly heading toward this when he defines pathology as a "simple extension going more or less beyond the higher or lower limits of variation proper to each phenomenon of the normal organism." But in the end it must be recognized that the terms used here, although only vaguely and loosely quantitative, still have a qualitative ring to them. Comte took from Broussais this vocabulary which fails to express what he wanted, and so it is to Broussais that we return in order to understand the uncertainties and gaps in Comte's exposition.

We prefer to base our summary of Broussais's theory on his treatise *De l'irritation et de la folie*, since, of all his works, this is the one Comte knew best. We have been able to determine that neither the *Traité de physiologie appliquée à la pathologie* [Treatise on Philosophy Applied to Pathology] nor the *Catéchisme de médecine physiologique* formulates this theory any more clearly or differently.[6] Broussais saw the vital primordial fact in excitation. Man exists only through the excitation exercised on his organs by the environment in which he is compelled to live. Through their innervation both the internal and external surfaces of contact transmit this excitation to the brain, which sends it back to all the tissues including the surfaces of contact. These surfaces are exposed to two kinds of excitation: foreign bodies and the influence of the brain. It is under the continuous action of these multiple sources of excitation that life is sustained. Applying the physiological doctrine to pathology means trying to find out how "this excitation can deviate from the normal state and constitute an abnormal or diseased state" [18, *263*]. These deviations are either deficiencies or excesses. Irritation differs from excitation only in terms of degree; it can be defined as the ensemble of disturbances "produced in the economy by agents which make vital phenomena more or less pronounced than they are in the normal state" [18, *267*]. Irritation is thus "normal excitation transformed by its excess" [18, *300*]. For example, through lack of oxygen, asphyxia deprives the lungs of its normal excitant. Inversely, air with too high an oxygen content "overexcites the lungs so much more strongly that the organ is more excitable and inflammation is the result" [18, *282*]. The two deviations, brought about by deficiency or excess, do not have the same importance in pathology, the latter considerably outweighing the former: "This second source of disease, the excess of excitation converted into irritation is thus much richer than the first, the lack of excitation, and it can be stated that most of our ills stem from this second source" [18, *286*]. In using them interchangeably, Broussais equates the terms *abnormal, pathological* and *morbid* [18, *263, 287, 315*]. The distinction between the normal or physiological and the abnormal or pathological would then be a simple quantitative one limited to the terms of deficiency and excess. And once Broussais admitted the physiological theory of the intellectual faculties, this distinction is valid for mental as well as organic phenomena [18, *440*]. This then, in summary, is the thesis whose fortune certainly owes more to the personality of the author than to the coherence of his text.

To begin with, in his definition of the pathological state, Broussais obviously confuses cause and effect. A cause can vary quantitatively so that it nevertheless both continues and provokes qualitatively different effects. To take a simple example, a quantitatively increased excitation can bring about a pleasant state, soon followed by pain, two feelings no one would want to confuse. In such a theory two points of view are being constantly mixed together, that of the sick man who is experiencing his illness and who is tested by it, and that of the scientist who finds nothing in disease that cannot be explained by physiology. But the states of an organism are like those found in music: the laws of acoustics are not broken in cacophony – this does not mean that all combinations of sounds are agreeable.

In short, such a conception can be developed in two slightly different directions, depending on whether the relation established between the normal and the pathological is one of *homogeneity* or continuity. Bégin, a strictly obedient disciple, adheres particularly to the relation of continuity:

> Pathology is no more than a branch, a result, a complement of physiology, or rather, physiology embraces the study of vital actions at all stages of the existence of living things. Without noticing, we pass from one to the other of these sciences as we examine functions from the moment the organs are performing with all the regularity and uniformity of which they are capable, to the point when the lesions are so serious that all functions become impossible and all movement stops. Physiology and pathology clarify each other [3, *XVIII*].

But it must be said that the continuity of a transition between one state and another can certainly be compatible with the heterogeneity of these states. The continuity of the middle stages does not rule out the diversity of the extremes. Broussais's own vocabulary sometimes betrays his difficulty in sustaining his assertion of a real homogeneity between normal and pathological phenomena; for example: "diseases increase, decrease, interrupt, *corrupt*[7] the innervation of the brain in terms of its instinctive, intellectual, sensitive and muscular relations" [18, *114*]; and: "the irritation which develops in living tissues does not always *alter*[8] them in a manner that constitutes inflammation" [18, *301*]. In the case of Comte, the vagueness of the notions of *excess* and *deficiency* and their implicit qualitative and normative character is even more noticeable, scarcely hidden under their metrical pretentions. Excess or deficiency exist in relation to a scale deemed valid and suitable – hence in relation to a norm. To define the abnormal as too much or too little is to recognize the normative character of the so-called normal state. This normal or physiological state is no longer simply a disposition which can be revealed and explained as a fact, but a manifestation of an attachment to some value. When Bégin defines the normal state as one where "the organs function with all the regularity and uniformity of which they are capable," we cannot fail to recognize that, despite Broussais's horror of all ontology, *an ideal of perfection soars over this attempt at a positive definition.*

From here on one can outline the major objection to this thesis according to which pathology is an extended or broadened physiology. The ambition to make pathology, and consequently therapeutics, completely scientific by simply making them derive from a previously established physiology would make sense only if, first, the normal could be defined in a purely objective way as a fact and second, all the differences between the normal state and the pathological state could be expressed in quantitative terms, for only quantity can take into account both homogeneity and variation. By questioning this double possibility we do not intend to undervalue either physiology or pathology. At any rate it must be evident that neither Broussais nor Comte fulfilled the two requirements which seem inseparable from the attempt with which their names are associated.

As far as Broussais is concerned this fact is not surprising. Methodical thinking was not his strength. For him the theses of physiological medicine were valuable less as speculative anticipation to justify painstaking research, than as a therapeutic prescription, in the form of bloodletting, to be imposed on everything and everyone. Armed with his lancet he aimed especially at inflammation found in the general phenomenon of excitation which had been transformed by its excess into irritation. As far as his teachings are concerned, their incoherence must be attributed to the fact that they embody, without too much care for their respective implications, the teachings of Xavier Bichat and John Brown, about whom it would be appropriate to say a few words.

First a student, then a rival of Cullen (1712–1780), the Scottish physician John Brown (1735–1788) had learned from his teacher about the notion of irritability suggested by Glisson (1596–1677) and developed by Haller. Author of the first great treatise on physiology (*Elementa physiologiae*, 1755–1766), Haller, a universal and gifted spirit, understood irritability to be the tendency of certain organs, particularly the muscles, to respond to any stimulus with a contraction. Contraction is not a mechanical phenomenon analogous to elasticity; it is the specific response of muscular tissue to different external stimuli [*sollicitations*]. By the same token, sensibility is the specific property of nervous tissue [29, *II*; 13 *bis, II*; 107, *51*; 110].

According to Brown, life is sustained by means of one particular property alone, excitability, which allows living organisms to be affected and to react. In the form of either *sthenia* or *asthenia*, diseases are simply a quantitative modification of this property wherever the excitation is either too strong or too weak.[9]

> It has been proved that health and disease are the same state depending upon the same cause, that is, excitement, varying only in degree; and that the powers producing both are the same, sometimes acting with a proper degree of force, at other times either with too much or too little; that the whole and sole province of a physician, is not to look for morbid states and remedies which have no existence, but to consider the deviation of excitement from the healthy standard, in order to remove it by the proper means (pp. 78–79).

Dismissing both the solidists and the humorists, Brown asserted that disease depends not on the primitive flaw of solids or fluids, but solely on the variations of the intensity of the excitation. Treating diseases means adjusting the amount of excitation to a greater or lesser degree. Charles Daremberg summarized these ideas in the following way:

> Brown took for his own and adapted to his own system a proposition I have called to your attention several times in these lectures, namely that pathology is a province of physiology, or as Broussais said, of pathological physiology. In fact, Brown asserts (§ 65) that it has been fully proven that the state of health and that of disease are not different, for the very reason that the forces which produce or destroy both have the same action; he tries to prove it, for example, by comparing muscle contraction and spasms or tetanus (§ 57 *et seq.*; cf. 136) [29, *1132*].

Without doubt what is particularly interesting in Brown's theory, as Daremberg notes repeatedly, is that it is the point of departure of Broussais's ideas, but even more interesting is the fact that to a certain degree it has a vague tendency to end up as a pathological phenomenon. Brown claimed to evaluate numerically the variable disposition of the organs to be excited:

> Suppose the greater affection of a part (as the inflammation of the lungs in petipneumony, the inflammation of the foot in the gout, the effusion of water into a general or particular cavity in dropsy) to be as 6, and the lesser affection of every other part to be 3, and the number of the parts less affected to amount to 1000; then it will follow, that the ratio of affection, confined to the part, to the affection of all the rest of the body, will be as 6 to 3000. This estimate, or something very like to it, is proved by the effect of the exciting hurtful powers, which always act upon the whole body; and by that of the remedies, which always remove the effect of the hurtful powers from the

> whole body, in every general disease (pp. 23–24).

Thereapeutics is based on calculation:

> Suppose the sthenic diathesis mounted up to 60 in the scale; to reduce it to 40 it is evident, that the 20 degrees of superfluous excitement must be taken off, and therefore, that remedies operating with a stimulus, weak enough to produce that effect, must be employed... (pp. 43–44 note).

Certainly we can and should smile at this caricature of the mathematization of pathological phenomena, but only on the condition that we agree that this doctrine does meet in full the demands of its postulates and that its concepts are completely coherent, something that is not true in Broussais.

Better still, a disciple of Brown, Samuel Lynch, in the same spirit constructed a scale of degrees of excitation, "a veritable thermometer of health and disease," as Daremberg called it, in the form of a proportional Table annexed to the various editions or translations of the *Elementa medicinae*. This table has two parallel scales from 0 to 80 going in opposite directions so that the maximum of excitability (80) corresponds to "0" of excitation and vice versa. Starting from perfect health (excitation = 40, excitability = 40) and going in both directions, the various degrees on the scale correspond to diseases, their causes, influences and treatments. For example, between 60 and 70 on the excitation scale are found the diseases of sthenic diathesis; peripneumonia, brain fever, severe smallpox, severe measles, severe erysipelas and rheumatism. For these the therapeutic indication is as follows: "In order to effect a cure, excitation must be decreased. This is achieved by avoiding overly strong stimuli, admitting only the weakest or negative stimuli. Cures are bloodletting, purging, diet, inner peace, cold, etc."

It must be said that this disinterment of an obsolete nosology was not intended to amuse or to satisfy the vain curiosity of a scholar. In a unique way it approaches a precise statement of the profound sense of the thesis now before us. Logically speaking, it is quite correct that an identification of phenomena, whose qualitative differences are considered illusory, takes the form of a quantification. Here the form of metrical identification is simply a caricature. But often a caricature reveals the essence of a form better than a faithful copy. It is true that Brown and Lynch succeeded only in constructing a conceptual hierarchy of pathological phenomena, a qualitative device to mark the state between the two extremes of health and illness. Marking is not measuring, a mark [*degré*] is not a cardinal unit. But even the error is instructive; it most certainly reveals the theoretical significance of one attempt, as well as the limits encountered in the object itself on which the attempt was made.[10*]

If we admit that Broussais was able to learn from Brown that, some quantitative variations apart, the assertion of the identity of normal and pathological phenomena logically means superimposing a system of measurement on research, what he learned from Bichat certainly counterbalanced that influence. In his *Recherches sur la vie et la mort* [Research on Life and Death] (1800), Bichat contrasts the object and methods of physiology with the object and methods of physics. According to him, instability and irregularity are the essential characteristics of vital phenomena, such that forcing them into a rigid framework of metrical relations distorts their nature [12, *art.* 7, § *I*]. It was from Bichat that Comte and even Claude Bernard took their systematic distrust of any mathematical treatment of biological facts, particularly any research dealing with averages and statistical calculations.

Bichat's hostility toward all metrical designs in biology was paradoxically allied with his assertion that diseases must be explained in terms of the definitely quantitative variations of their properties, with the tissues which make up the organs serving as a scale.

> To analyse precisely the properties of living bodies; to show that every physiological phenomenon is, in the final analysis, related to these properties considered in their natural state and that every pathological phenomenon derives from their increase, decrease, or alteration, that every therapeutic phenomenon has as its principle the return to the natural type from which they had deviated; to determine precisely the cases where each one comes into play . . . this is the general idea of this work [13, *I, XIX*].

Here is the source of that ambiguity of ideas which we have already criticized in Broussais and Comte. Augmentation and diminution are concepts which connote quantity, but alteration is a concept of qualitative force. One cannot, of course, blame physiologists and physicians for falling into that trap of the Same and the Other into which so many philosophers have fallen since Plato. But it is good to be able to recognize the trap and not blithely ignore it just when one is caught. All of Broussais's teachings are contained in embryo in this proposition of Bichat:

> All curative resources have only one goal, to return altered vital properties to their natural state. All means which fail to diminish the increased organic sensibility in inflammation, which do not increase the completely diminished property in edemas, infiltration, etc., which do not lower animal contractility in convulsions and do not raise it in paralysis, etc., essentially miss their goal; they are contra-indicated [13, *I*, *12*].

The only difference is that Broussais reduced all pathogeny to a phenomenon of increase and excess and, consequently, all therapy to bloodletting. Here it is certainly true to say that excess in everything is a defect!

It may be surprising to see that an exposition of Comte's theory has turned into a pretext for a retrospective study. Why wasn't a chronological order employed at the outset? Because a historical narrative always reverses the true order of interest and inquiry. It is in the present that problems provoke reflection. And if reflection leads to a regression, the regression is necessarily related to it. Thus the historical origin is really less important than the reflective origin. Certainly Bichat, the founder of histology, owes nothing to Comte. It is not even certain that the resistance encountered by the cellular theory in France is really broadly related to Charles Robin's positivist loyalties. It is known that Comte, following Bichat, did not admit that analysis could go beyond tissues [64]. What is certain in any case is that even in the milieu of medical culture, the theories of general pathology originated by Bichat, Brown and Broussais were influential only to the extent that Comte found them advantageous. The physicians of the second half of the nineteenth century were for the most part ignorant of Broussais and Brown, but few were unaware of Comte or Littré; just as today most physiologists cannot ignore Bernard, but disregard Bichat to whom Bernard is connected through Magendie.

By going back to the more remote sources of Comte's ideas – through the pathology of Broussais, Brown and Bichat – we put ourselves in a better position to understand their significance and limits. We know that it was from Bichat (through the intermediary of his teacher in physiology, de Blainville) that Comte acquired a decided hostility toward all mathematization of biology. He accounts for this at great length in the fortieth lecture of the *Cours de philosophie positive*. That influence of Bichat's vitalism on the Comtean positivist conception of vital phenomena, however discreet, balances the profound logical requirements of the assertion of the identity between physiological and pathological mechanisms, requirements moreover ignored by Broussais, another intermediary between Comte and Bichat on one precise point of pathological doctrine.

One must bear in mind that Comte's aims and intentions are very different from Broussais's, or rather, different from Broussais's intellectual antecedents, when he develops the same conceptions in pathology. On the one hand, Comte claims to be codifying scientific methods, on the other, to be establishing a political doctrine scientifically. By stating in a general way that diseases do not change vital phenomena, Comte is justified in stating that the cure for political crises consists in bringing societies back to their essential and permanent structure, and tolerating progress only within limits of variation of the natural order defined by social statics. In positivist doctrine, Broussais's principle remains an idea subordinated to a system, and it is the physicians, psychologists and men of letters, positivist by inspiration and tradition, who disseminated it as an independent conception.

CHAPTER III

Claude Bernard and Experimental Pathology

It is certain that Claude Bernard never referred to Comte while dealing with the problem of the relationship between the normal and the pathological, although he did solve it in an apparently similar fashion; it is equally certain that he could not ignore Comte's opinions. We know that Claude Bernard read Comte closely, and with pen in hand, as borne out by notes dating probably from 1865–66, and published in 1938 by Jacques Chevalier [11]. For the physicians and biologists of the Second Empire, Magendie, Comte and Claude Bernard are three gods – or three devils – of the same religion. In examining the experimental work of Bernard's teacher, Magendie, Littré analyzes those postulates which coincide with Comte's ideas on experimentation in biology and its relation to the observation of pathological phenomena [78, *162*]. E. Gley was the first to show that Claude Bernard, in his article "Progrès des sciences physiologiques" *(Revue des Deux Mondes*, 1 August 1865), took for his own the law of the three states, and that he had a part in publications and associations in which Charles Robin made the positivist influence felt [44, *164–170*]. In 1864, together with Brown-Séquard, Robin published the *Journal de l'anatomie et de la physiologie normales et pathologiques de l'homme et des animaux*: reports of Bernard, Chevreul, etc. appeared in the first issues. Bernard was the second president of the Société de Biologie which Robin had founded in 1848, whose guiding principles were formulated in a lecture to the charter members:

> By studying anatomy and the classification of living beings, we hope to clarify the mechanism of functions; by studying physiology, to come to know how organs can be changed and within what limits functions deviate from the normal [44, *166*].

For his part, Lamy has shown that, in practice, nineteenth-century artists and writers, who looked for sources of inspiration or themes to reflect upon in physiology and medicine, did not distinguish between the ideas of Comte and those of Bernard [68].

Having said that, we must add that it is really a very difficult and delicate task to outline Claude Bernard's ideas on the precise problem of the nature and meaning of pathological phenomena. Here is a scientist of note whose discoveries and methods still bear fruit today, to whom physicians and biologists refer constantly, and for whose works there is no complete critical edition! Most of the lectures given at the Collège de France were edited and published by students. But that which Bernard himself did write, his correspondence, has not been the object of any fair, methodical investigation. Notes and notebooks have been published here and there and have immediately become the center of controversy so expressly tendentious that one wonders whether the same insinuations, which are moreover quite varied, did not actually provoke the publication of all these fragments. Bernard's thought remains a problem. The only honest solution will be the systematic publication of his papers and, when this decision is finally reached, the placing of his papers in an archive.[11]

In Bernard's work, the real identity – should one say in mechanisms or symptoms or both? – and continuity of pathological phenomena and the corresponding physiological phenomena are more a monotonous repetition than a theme. This assertion is to be found in the *Leçons de physiologie expérimentale appliquée à la médecine* [Lectures on Experimental Physiology Applied to Medicine] (1855), especially in the second and twenty-second lectures of Vol. II, and in the *Leçons sur la chaleur animale* [Lectures on Animal Heat] (1876). We prefer to choose the *Leçons sur le diabète et la glycogenèse animale* [Lectures on Diabetes and Animal Glycogenesis] (1877) as the basic text, which, of all Bernard's works, can be considered the one especially devoted to illustrating the theory, the one where clinical and experimental facts are presented at least as much for the "moral" of a methodological and philosophical order which can be drawn from it, as for their intrinsic physiological meaning.

Bernard considered medicine as the science of diseases, physiology as the science of life. In the sciences it is theory which illuminates and dominates practice. Rational therapeutics can be sustained only by a scientific pathology, and a scientific pathology must be based on physiological science. Diabetes is one disease which poses problems whose solution proves the preceding thesis.

> Common sense shows that if we are thoroughly acquainted with a physiological phenomenon, we should be in a position to account for all the disturbances to which it is susceptible in the pathological state: Physiology and pathology are intermingled and are essentially one and the same thing [9, 56].

Diabetes is a disease which consists solely and entirely in the disorder of a normal function.

> Every disease has a corresponding normal function of which it is only the disturbed, exaggerated, diminished or obliterated expression. If we are unable to explain all manifestations of disease today, it is because physiology is not yet sufficiently advanced and there are still many normal functions unknown to us [9, *56*].

In this Bernard was opposed to many physiologists of his day, according to whom disease was an extra-physiological entity, superimposed on the organism. The study of diabetes no longer allowed such an opinion.

> In effect diabetes is characterized by the following symptoms: polyuria, polydipsia, polyphagia, autophagia, and glycosuria. Strictly speaking, none of these symptoms represents a new phenomenon, unknown to the normal state, nor is any a spontaneous production of nature. On the contrary all of them preexist, save for their intensity which varies in the normal state and in the diseased state [9, *65–66*].

It is easy to demonstrate this as far as polyuria, polydipsia, polyphagia and autophagia are concerned, less easy with regard to glycosuria. But Bernard contended that glycosuria is a "masked and unnoticed" phenomenon in the normal state and that only its exaggeration makes it apparent [9, *67*]. In reality Bernard does not effectively prove what he is propounding. In the sixteenth lecture, after comparing the opinions of physiologists, who assert the constant presence of sugar in normal urine, with that of those who deny it, after having shown the difficulty of experiments and of their control, Bernard adds that in the normal urine of an animal fed on nitrogenized substances and deprived of sugar and starches, he never succeeded in uncovering the faintest traces of sugar, but that it would be completely different with an animal fed on excessive amounts of sugar and starches. It is equally natural to think, he says, that in the course of its oscillations, glycemia can determine the passage of sugar in the urine.

> In sum, I do not believe that this proposition can be formulated as an absolute truth: there is sugar in normal urine. But I readily admit that there are many, many cases where there are traces; there is a kind of transient glycosuria which here as everywhere establishes an imperceptible and elusive passage between the physiological and the pathological states. I agree in other respects with clinicians in recognizing that the glycosuric phenomenon has no real, well established pathological character until it becomes permanent [9, *390*].

It is striking to document here that, in trying to furnish a particularly convincing fact favoring his interpretation in a case where he felt especially challenged, Bernard found himself forced to admit this same fact without experimental proof – by reason of the theory – by supposing that its reality was situated beyond the limits of sensibility of all the methods used at that time for its detection. Today H. Frédéricq admits on this very point that there is no normal glycosuria, that in certain cases where a large amount of liquid is ingested and there is copious diuresis, glucose cannot be reabsorbed in the kidney at the level of the convoluted tube and is, so to speak, washed away [40, *353*]. This explains why certain writers like Nolf can say that there is a normal infinitesimal glycosuria [90, *251*]. If there is no glycosuria normally, what physiological phenomenon does diabetic glycosuria exaggerate quantitatively?

Briefly, we know that Claude Bernard's genius lies in the fact that he showed that the sugar found in an animal organism is a product of this same organism and not just something introduced from the plant world through its feeding; that blood normally contains sugar, and that urinary sugar is a product generally eliminated by the kidneys when the rate of glycemia reaches a certain threshold. In other words, glycemia is a constant phenomenon independent of food intake to such an extent that it is the absence of blood sugar that is abnormal, and glycosuria is the consequence of glycemia which has risen above a certain quantity, serving as a threshold. In a diabetic, glycemia is not in itself a pathological phenomenon – it is so only in terms of its quantity; in itself, glycemia is a "normal and constant phenomenon in a healthy organism" [9, *181*].

> There is only one glycemia, it is constant, permanent, both during diabetes and outside that morbid state. Only it has degrees: glycemia below 3 to 4% does not lead to glycosuria; but above that level glycosuria results.... It is impossible to perceive the transition from the normal to the pathological state, and no problem shows better than diabetes the intimate fusion of physiology and pathology [9, *132*].

The energy Bernard spent expounding his thesis does not seem superfluous if the thesis is placed in a historical perspective. In 1866 Jaccoud, professor *agregé* at the Faculty of Medicine in Paris, dealt with diabetes in a clinical lecture by saying that glycemia is an inconstant, pathological phenomenon and that the production of sugar in the liver is, according to the work of Pavy, a pathological phenomenon.

> The diabetic state cannot be attributed to the overintensification of a physiological operation which does not exist.... It is impossible to regard diabetes as the overintensification of a reg-

> ular operation: it is the expression of an operation completely foreign to normal life. This operation is in itself the essence of the disease [57, *826*].

In 1883, when Bernard's theory was more firmly established, Jaccoud, by then professor of internal pathology, continued to maintain his objections in his *Traité de pathologie interne* [Treatise on Internal Pathology]: "The transformation of glycogen into sugar is either a pathological or cadaverous phenomenon" [58, *945*].

If we really want to understand the meaning and significance of the assertion of continuity between normal and pathological phenomena, we must bear in mind that the thesis toward which Bernard's critical demonstrations are directed is one which admits a qualitative difference between the mechanisms and products of the vital functions of the normal state and those of the pathological state. This contradiction of thesis appears perhaps more clearly in the *Leçons sur la chaleur animale*:

> Health and disease are not two essentially different modes as the ancient physicians believed and some practitioners still believe. They should not be made into distinct principles, entities which fight over the living organism and make it the theater of their contest. These are obsolete medical ideas. In reality, between these two modes of being, there are only differences of degree: exaggeration, disproportion, discordance of normal phenomena constitute the diseased state. There is no case where disease would have produced new conditions, a complete change of scene, some new and special products [8, *391*].

To support this Bernard gives an example which he believes particularly suited to ridicule the opinion he is fighting. After two Italian physiologists, Lussana and Ambrossoli, repeated his experiments on the cutting of the sympathetic nerve and its effects, they denied the physiological character of the heat engendered by the vasodilatation of the effected organs. According to them, this heat was morbid, different in every respect from physiological heat, the latter originating from the combustion of food, the former from the combustion of tissues. As if food, Bernard replied, were not always burned at the level of tissues of which it becomes an integral part. Thinking that he had easily refuted the Italian writers, Bernard added:

> In reality, physico-chemical manifestations do not change in nature, whether they take place inside or outside the organism, in a healthy or diseased state. There is only one kind of calorific agent; whether it is produced in a furnace or in an organism it is none the less the same. There cannot be physical heat and animal heat, still less, morbid heat and physiological heat. Morbid animal heat and physiological heat differ only in degree, not in their nature [8, *394*].

Hence the conclusion:

> These ideas of a struggle between two opposing agents, of antagonism between life and death, between health and sickness, inanimate and living nature have had their day. The continuity of phenomena, their imperceptible gradation and harmony must be recognized everywhere [*ibid.*].

These last two texts seem to me to be particularly illuminating because they reveal a chain of ideas found nowhere in the *Leçons sur le diabète*. The idea of the continuity between the normal and the pathological is itself in continuity with the idea of the continuity between life and death, organic and inorganic matter. Bernard has the indisputable merit of having denied the antithesis admitted until then between the organic and the mineral, plant and animal, of having affirmed the universal applicability of the determinist postulate and the material identity of all physicochemical phenomena regardless of their setting and appearance. He was not the first to assert the identity of the chemical products of the laboratory and those of "living" chemistry – that idea was conceived after Wöhler succeeded in synthesizing urea in 1828 – he simply "reinforced the physiological impulse given organic chemistry by the works of Dumas and Liebig."[12] But he was the first to assert the physiological identity of plant functions and corresponding animal functions. Until his time it was held that plant respiration was the inverse of that of animals, that plants fixed carbon and animals burned it, that plants performed reductions and animals combustions, that plants produced syntheses which animals destroyed by using them, as they were incapable of producing anything similar.

Bernard denied all of these antitheses, and the discovery of the glycogenic function of the liver is one of the most beautiful results of the desire to "recognize everywhere the continuity of phenomena."

One probably does not have to ask now whether Bernard formed a correct idea of what constitutes an antithesis or contrast, and whether it is justifiable to consider the pair of notions, health–disease as symmetrical with the pair life–death, to draw the conclusion that once he identified the terms of the second, he was authorized to seek the identification of the terms of the first. One will probably ask what Bernard meant by asserting the unity of life and death. For the purposes of lay or religious polemic, it is often asked whether Bernard was a materialist or a vitalist.[13] It seems that a careful reading of the *Leçons sur les phénomènes de la vie* [Lectures on the Phenomena of Life] (1878) suggests an answer full of nuances. From the physicochemical point of view, Bernard did not accept the distinction between the organic realm and the mineral realm: "The chemistry of the laboratory and the chemistry of life are subject to the same laws: *there are not two chemistries*" [10, *I, 224*]. This amounts to saying that scientific analysis and experimental techniques can identify and reproduce products of vital syntheses as well as inorganic objects. But this simply asserts the homogeneity of matter within the living form and outside of this form, for in refusing mechanistic materialism, Bernard asserts the originality of the living form and its functional activities:

> Although the vital manifestations are placed under the direct influence of physico-chemical conditions, these conditions cannot organize, harmonize phenomena in the order and succession which they assume particularly in living things [10, *II, 218*].

And still more precisely:

> Along with Lavoisier I believe that living things are tributaries of the general laws of nature and that their manifestations are physical and chemical expressions. Unlike physicists and chemists I am far from seeing vital actions in the phenomena of the inanimate world – on the contrary I believe that the expression is particular, the mechanism special, the agent specific although the result is the same. No chemical phenomenon exists inside the body as it does outside of it [*ibid.*].

These last words could serve as an epigraph for the work of Jacques Duclaux on the *Analyse physico-chimique des fonctions vitales* [Physico-chemical Analysis of Vital Functions]. According to Duclaux, who, in this work was obviously far from any kind of spiritualism, no intracellular chemical reaction can be represented by an equation derived from experimentation *in vitro*:

> As soon as a body can be represented by our symbols, living matter considers it an enemy and eliminates or neutralizes it.... Man has created a chemistry which has developed from natural chemistry without being confused with it [36].

Be that as it may, it seems clear that for Bernard, recognizing the continuity of phenomena does not mean ignoring their originality. Given this, and keeping the symmetry, could one not say what he says of the relations between inanimate and living matter? – there is only one physiology, but far from seeing the type of pathological phenomena in physiological phenomena, one must consider that its expression is particular, its mechanism special, although the result is identical; no phenomenon exists in the diseased organism as it does in the healthy one. Why assert unreservedly the identity of disease and health when one does not do so for life and death, when one intends to use the relation between the latter as a model for that between the former?

Claude Bernard, unlike Broussais and Comte, supported his general principle of pathology with verifiable arguments, protocols of experiments and above all methods for quantifying physiological concepts. Glycogenesis, glycemia, glycosuria, combustion of food, heat from vasodilatation are not qualitative concepts but the summaries of results obtained in terms of measurement. From here on we know exactly what is meant when it is claimed that disease is the exaggerated or diminished expression of a normal function. Or at least we have the means to know it, for in spite of Bernard's undeniable progress in logical precision, his thought is not entirely free from ambiguity.

First of all, with Bernard as with Bichat, Broussais and Comte, there is a deceptive mingling of quantitative and qualitative concepts in the given definition of pathological phenomena. Sometimes the pathological state is "the disturbance of a normal mechanism consisting in a quantitative variation, an exaggeration or attenuation of normal phenomena" [9, *360*], sometimes the diseased state is made up of "the exaggeration, disproportion, discordance of normal phenomena" [8, *391*]. Who doesn't see that the term "exaggeration" has a distinctly quantitative sense in the first definition and a rather qualitative one in the second? Did Bernard believe that he was eradicating the qualitative value of the term "pathological" by substituting for it the terms dis-turbance, dis-proportion, dis-cordance?

This ambiguity is certainly instructive in that it reveals that the problem itself persists at the heart of the solution presumably given to it. And the problem is the following: is the concept of disease a concept of an objective reality accessible to quantitative scientific knowledge? Is the difference in value, which the living being establishes between his normal life and his pathological life, an illusory appearance which the scientist has the legitimate obligation to deny? If this annulling of a qualitative contrast is theoretically possible, it is clear that it is legitimate; if it is not possible, the question of its legitimacy is superfluous.

It has been pointed out that Bernard uses two expressions interchangeably, *quantitative variations* and *differences of degree*, that is, he makes two concepts of them, *homogeneity* and *continuity*, the first used implicitly, the second, expressly. The use of either of these concepts does not entail the same logical requirements. If I assert the homogeneity of two objects, I must at least define the nature of one of the two or rather some nature common to both. But if I assert a continuity, I can only interpolate between the two extremes all the intermediaries at my disposal, without reducing one to the other, by divisions of progressively smaller intervals. This is so true that certain writers claim continuity between health and disease in order to refuse to define either of them.[14] They say that there is no completely normal state, no perfect health. This can mean that there exist only sick men. In an amusing way Molière and Jules Romains have shown what kind of "iatrocracy" can justify this assertion. But this could also mean that there are no sick men, which is nonetheless absurd. One wonders whether physicians, in stating seriously that perfect health does not exist and that consequently disease cannot be defined, have suspected that they were purely and simply reviving the problem of the existence of the perfect and the ontological argument.

For a long time people tried to find out whether they could prove the existence of the perfect being starting with its quality of perfection, since, having all the perfections, it would also have that of bringing about its own existence. The problem of the actual existence of perfect health is analogous. As if perfect health were not a normative concept, an ideal type? Strictly speaking a norm does not exist, it plays its role which is to devalue existence by allowing its correction. To say that perfect health does not exist is simply saying that the concept of health is not one of an existence, but of a norm whose function and value is to be brought into contact with existence in order to stimulate modification. This does not mean that health is an empty concept.

But Claude Bernard is far from such a facile relativism, owing to the fact that first, the assertion of continuity in his thought always implies that of homogeneity, and second, he thinks that it is always possible to give an experimental content to the concept of the normal. For example, what he calls an animal's normal urine is the urine of an animal with an empty stomach, always comparable to itself – the animal feeding itself in the same way with its own reserves – and such that it serves as a constant frame of reference for all the urine obtained in the feeding conditions which he wants to set up [5, *II, 13*]. Later on we will discuss the relations between the normal and the experimental. Right now, we only want to examine Bernard's point of view when he conceives of the pathological phenomenon as a quantitative variation of the normal phenomenon. Naturally it is understood that if in the course of this examination we use recent physiological or clinical data, it is not to reproach Bernard for having ignored what he could not know.

If glycosuria is considered to be the major symptom of diabetes, the presence of sugar in diabetic urine makes it qualitatively different from normal urine. In terms of the physiological state, the pathological state, when identified with its principal symptom, is a new quality. But if in considering urine as a product of renal secretion, the physician's thought turns to the kidney and the relationship between the renal filter and the composition of the blood, he will consider glycosuria as excess glycemia pouring over a threshold. The glucose overflowing the threshold is qualitatively the same as the glucose normally held back by the threshold. The only difference is, in effect, one of quantity. If, then, the renal mechanism of urinary secretion is considered in terms of its results – physiological effects or morbid symptoms – disease is the appearance of a new quality; if the mechanism is considered in itself, disease is only a quantitative variation. Likewise, alkaptonuria could be cited as an example of a normal chemical mechanism capable of producing an abnormal symptom. Discovered by Boedeker in 1857, this rare disease consists essentially in a disturbance of the metabolism of an amino acid, tyrosine. Alkaptone or homogentisic acid is a normal product of the intermediate metabolism of tyrosine, but alkaptonuric diseases are distinguished by their incapacity to go beyond this phase and burn homogentisic acid [41, *10.534*]. Homogentisic acid then passes into the urine and is transformed in the presence of alkalis through oxidation to give off a black pigment coloring the urine and giving it a new quality which is in no way an exaggeration of some quality present in normal urine. Moreover, alkaptonuria can be brought about experimentally by a massive absorption of tyrosine (50 g every 24 hours). Thus, we have a pathological phenomenon which can be defined in terms of quality or quantity depending on one's point of view, depending on whether the vital phenomenon is considered in terms of its expression or its mechanism.

But can one choose one's point of view? Is it not obvious that if we want to work out a scientific pathology we must consider real causes and not apparent effects, functional mechanisms and not their symptomatic expressions? Is it not obvious that by relating glycosuria to glycemia and glycemia to hepatic glycogenesis, Bernard was considering the mechanisms, the scientific explanation of which derives from a number of quantitative relations; for example, the physical laws of the equilibria of membranes, the law of the concentration of solutions, the reactions of organic chemistry, etc.?

All of this would be indisputable if physiological functions could be considered as mechanisms, thresholds as barriers, regulations as safety valves, servo-brakes or thermostats. Are we about to fall into all the traps and dangers of the iatro-mechanist conceptions? To take the same example of diabetes, today we are far from thinking that glycosuria is only a function of glycemia and that the kidney simply prevents the filtration of glucose by means of a constant threshold (of 1.70 pph and not 3 pph as Bernard first thought). According to Chabanier and Lobo-Onell: "The renal threshold is essentially *mobile*, and its *behavior*, *variable*, depending on the patients" [25, *16*]. On the one hand, in subjects without hyperglycemia, glycosuria can sometimes be demonstrated, even higher than that of true diabetics. This is spoken of as renal glycosuria. On the other hand, in subjects whose glycemia sometimes reaches 3 g and more, glycosuria can be practically nil. This is called pure hyperglycemia. Furthermore, two diabetics situated in the same conditions for observation and showing the same glycemia of 2.50 g in the morning on an empty stomach, can show a variable glycosuria, one losing 20 g and the other, 200 g of glucose in their urine [25, *18*].

We are now led to modify the classic scheme, which linked glycosuria to basal disturbance by the sole intermediary of hyperglycemia, by introducing a new articulation between hyperglycemia and glycosuria: "*renal behavior*" [25, *19*]. By speaking of the mobility of the threshold, of renal behavior, a notion is introduced in the explanation of the mechanism of urinary secretion that cannot be entirely translated into analytical and quantitative terms. This amounts to saying that to become a diabetic is to change kidneys, a proposition which will seem absurd only to those who identify a function with its anatomical position. It seems permissible to conclude that by substituting mechanisms for symptoms in the comparison between the physiological and the pathological state, no difference in quality between the two states is eliminated at all.

This conclusion looms larger still when we stop dividing disease into a multiplicity of functional mechanisms gone wrong, and regard it as an event involving the living organism taken as a whole. This is very much the case of diabetes. Today we say it is a "diminution of the ability to use glucose in terms of glycemia" [25, *12*]. Von Mering and Minkowski's discovery in 1889 of *experimental pancreatic diabetes*, Laguesse's discovery of the endocrine pancreas, Banting and Best's isolation in 1920 of the insulin secreted by the islands of Langerhans, made possible the assertion that the fundamental disturbance in diabetes is hypoinsulinemia [diabetes mellitus]. Must it be said then that these researches, unsuspected by Bernard, finally confirmed his principles of general pathology? Certainly not, for in 1930–31 Houssay and Biasotti showed, by destroying both the pancreas and the pituitary in the toad and dog, that the roles of the pituitary and the pancreas were antagonistic in metabolism. Following total removal of the pancreas a healthy dog cannot survive for more than four or five weeks. But a combination of a hypophysectomy [removal of the pituitary] and a pancreatectomy produces considerable improvement in diabetes: glycosuria is very much reduced and, on an empty stomach, even suppressed; polyuria is suppressed, glycemia is near normal and weight loss is very much slowed down. Hence it seemed warranted to conclude that the action of insulin in the metabolism of glucides is not direct since diabetes can be lessened without the administration of insulin. In 1937 Young established that with an injection of an extract of the anterior lobe of the pituitary every day for about three weeks, a normal dog could sometimes be made definitely diabetic. Louis Hédon and Auguste Loubatières, who took up Young's study of experimental diabetes in France, concluded: "*Temporary* hyperactivity of the anterior lobe of the pituitary can be at the source of not only a transitory disturbance of glycoregulation but also *permanent diabetes* which persists indefinitely after the disappearance of the cause which set it off" [54, *105*]. Have we been sent from diminution to augmentation, and is Bernard's insight flawless just when we believed it at fault? It does not seem so because, all things considered, this pituitary hypersecretion is only a symptom, at the glandular level, of either a pituitary tumor or a general endocrinal readjustment (puberty, menopause, pregnancy). As far as internal secretions are concerned, as in the case of the nervous system, localizations are "privileged" rather than absolute and what appears to be partial augmentation or diminution is in fact an alteration in the whole. "Nothing is more illusory," writes Rathery,

> than to consider the metabolism of glucides as under the sole control of the pancreas and its secretion. The metabolism of glucides depends on many factors: (a) blood vascular glands; (b) the

liver; (c) the nervous system; (d) vitamins; (e) mineral elements, etc. Now any of these factors can come into play to bring about diabetes [98, *22*].

If we consider diabetes as a nutritional disease and constant glycemia as a tonus indispensable to the existence of the organism taken as a whole (Soula),[15] we are far from being able to draw the conclusions about general pathology from the study of diabetes that Claude Bernard drew from it in 1877.

These conclusions are to be criticized not so much for being wrong as for being inadequate and incomplete. They stem from the unwarranted extrapolation of a perhaps privileged case and, moreover, from a definition which is clumsy in terms of the point of view adopted. It is correct that certain symptoms are the quantitatively varied product of constant mechanisms of the physiological state. This would be the case, for example, with hyperchlorhydria in the ulcerous stomach. It is possible for some mechanisms to be the same in the state of health and in the state of disease. In the case of a stomach ulcer, the reflex which determines the secretion of gastric juices always seems to originate from the pyloric cavity, if it is true that they are stenosal ulcers near the pylorus accompanied by the most significant hypersecretion and if the removal of this region through a gastrectomy is followed by a reduction of the secretion.

But first of all, as far as the precise case of ulcers is concerned, it must be said that the essence of the disease consists not in hyperchlorhydria, but rather in the fact that here the stomach is digesting itself, a state which everyone will undoubtedly agree differs profoundly from the normal. Incidentally, perhaps this would be a good example to explain what a normal function is. A function could be said to be normal as long as it is independent of the effect it produces. The stomach is normal as long as it digests without digesting itself. What is true of balance scales is also true of functions: fidelity first, then sensitivity.

Furthermore, it must be said that the reduction of all pathological cases to the explanatory scheme proposed by Bernard is very remote. This is particularly true of the scheme put forward in the *Leçons sur la chaleur animale.* Of course there is no normal heat and pathological heat, in the sense that both can be expressed in terms of identical physical effects: the dilatation of a column of mercury in the course of taking a rectal or axillary temperature. But the identity of the heat does not involve the identity of the source of heat nor even the identity of the mechanism for liberating the calories. Claude Bernard answered his Italian adversaries by saying that animal heat always derives from food burned at the tissue level. But the same food can be burned in any number of ways, its breakdown stopping at different stages. To postulate, with reason, the identity of chemical and physical laws with one another, does not oblige one to ignore the specificity of the phenomena which reveal them. When in the course of measurement of basal metabolism, a woman suffering from Basedow's [or Graves's] disease breathes into a closed space whose variation in volume will give the rate of oxygen consumption, oxygen is always burned according to the chemical laws of oxidation (5 calories for one liter of O_2), and it is precisely by setting up the constancy of these laws in this case that one can calculate the variation in metabolism and term it abnormal. It is in this precise sense that there is an identity of the physiological and the pathological. But it could also be said that there is an identity of the chemical and the pathological. It will be agreed that this is one way to make the pathological disappear and not to clarify it. Isn't this also true of the case where it is declared homogeneous with the physiological?

By way of summary, Claude Bernard's theory is valid in certain limited cases:

1. as long as the pathological phenomenon is limited to some symptom, *leaving aside its clinical context* (hyperchlorhydria, hyperthermia or hypothermia; reflex hyperexcitability);
2. as long as symptomatic effects are traced back to *partial* functional mechanisms (glycosuria in terms of hyperglycemia; alkaptonuria in terms of the incomplete metabolism of tyrosine).

Even when limited to these precise cases, his theory runs into many difficulties. Who would maintain that hypertension is a simple increase in the physiological arterial pressure and neglect the profound alteration in the structure and function of the vital organs (heart and blood vessels, kidneys, lungs), an alteration such that it constitutes a new way of life for the organism, new behavior which prudent therapy must take into account by not treating the tension at an unpropitious moment in order to bring it back to the norm? Who would maintain that hypersensitivity to certain toxic substances is a simple quantitative modification of a normal reactivity, without first asking himself whether there isn't only the appearance (of the fact of poor renal elimination or of an overly rapid reabsorption in relation to a general defined state), without subsequently distinguishing isotoxic intolerance where phenomena are changed only quantitatively, and heterotoxic intolerance, where new symptoms appear in relation to a change of the cellular reactivity to the poison (A. Schwartz)?[16] The same is true of functional mechanisms, which can be easily experimented with separately. But in the living organism all functions are interdependent and their rhythms are coordinated: renal behavior can be only theoretically divorced from the behavior of the organism functioning as a whole.

By taking examples of the order of metabolic phenomena (diabetes, animal heat), Bernard found cases which were too unilateral to be generalized without some arbitrariness. How can infectious diseases, whose etiology and pathogenesis were then beginning to emerge from their prescientific borders, be explained within the framework of his ideas? Certainly the theory of inconspicuous [*inapparent*] infections (Charles Nicolle)[17] and the theory of *terrain* allow the assertion that infectious disease pushed roots into the so-called normal state. But this widespread opinion is not unassailable for all that. It is not normal for a healthy subject to have diphtheria bacilli lodged in his throat, in the same sense that it is normal for him to eliminate phosphates in his urine or contract his pupils when passing quickly from the dark into the light. A disease in a state of suspension or remission is not a normal state analogous to the exercising of a function, whose blockage would be fatal. Similarly, if it is a good idea to bear in mind the *terrain* as Pasteur himself advised, one should perhaps still not go to the length of making a microbe an epiphenomenon. It takes one last fragment of crystal to obtain the solidification of a supersatured solution. Strictly speaking, it takes a microbe to make an infection. Without doubt it has been possible to produce lesions like those of pneumonia or typhoid by means of physical or chemical irritation of the splanchnic nerve [80]. But in order to keep to the classical explanation of infection, one can try, once infection has occurred, to reestablish a certain continuity between before and after by using etiological antecedents. It seems difficult to assert that the infectious state produces no real discontinuity in the history of the living being.

Nervous diseases constitute another awkward fact for Bernard's explanation based on his principles. These have long been described in terms of exaggeration and deficiency. When the higher functions of life as it relates to the external world were considered as the sums of elementary reflexes, and the brain centers as pigeonholes for images or impressions, a quantitative explanation of pathological phenomena was inevitable. But the conceptions of Hughlings Jackson, Head and Sherrington, paving the way for more recent theories such as those of Goldstein, oriented research in directions where facts took on a synthetic qualitative value, at first ignored. We will come back to this later. It will be enough to say here briefly that according to Goldstein, normal behavior in relation to language disturbances, can be explained in pathological terms only on the condition that the notion of the modification of personality by disease is introduced. In general, any one act of a normal subject must not be related to an analogous act of a sick person without understanding the sense and value of the pathological act for the possibilities of existence of the modified organism:

> One must refrain from thinking that the different attitudes possible in a sick person merely represent a kind of residue of normal behavior, what survived destruction. The attitudes which have survived in the sick person never turn up in that form in a normal subject, not even in the inferior stages of its ontogenesis or phylogenesis, as it is all too frequently admitted. Disease has given them particular forms and they cannot be understood well unless the morbid state is taken into account [45, *437*].

In short, the continuity of the normal state and the pathological state does not seem real in the case of infectious diseases, no more than homogeneity in the case of nervous diseases.

By way of summary, in the medical domain, Claude Bernard, with the authority of every innovator who proves movement by marching, formulated the profound need of an era which believed in the omnipotence of a technology founded on science, and which felt comfortable in life in spite, or perhaps because of, romantic lamentations. An art of living – as medicine is in the full sense of the word – implies a science of life. Efficient therapeutics assumes experimental pathology, which in turn cannot be separated from physiology. "Physiology and pathology are intermingled and are one and the same thing." But must it be deduced from this, with brutal simplicity, that life is the same in health and disease, that it learns nothing in disease and through it? The science of opposites is one, said Aristotle. Must it be concluded from this that opposites are not opposites? That the science of life should take so-called normal and so-called pathological phenomena as objects of the same theoretical importance, susceptible of reciprocal clarification in order to make itself fit to meet the totality of the vicissitudes of life in all its aspects, is urgent far more than it is legitimate. This does not mean that pathology is nothing other than physiology, and still less that disease, as it relates to the normal state, represents only an increase or a reduction. It is understood that medicine needs an objective pathology, but research which causes its object to vanish is not objective. One can deny that disease is a kind of violation of the organism and consider it as an event which the organism creates through some trick of its permanent functions, without denying that the trick is new. An organism's behavior can be in continuity with previous behaviors and still be another behavior. The progressiveness of an advent does not exclude the originality of an event. The fact that a pathological symptom, considered by itself, expresses the hyperactivity of a function whose product is exactly identical with the product of the same function in so-called normal conditions, does not mean that an organic disturbance, conceived as another aspect of the whole of functional totality and not as a summary of symptoms, is not a new mode of behavior for the organism relative to its environment.

In the final analysis, would it not be appropriate to say that the pathological can be distinguished as such, that is, as an alteration of the normal state, only at the level of organic totality, and when it concerns man, at the level of conscious individual totality, where disease becomes a kind of evil? To be sick means that a man really lives another life, even in the biological sense of the word. To return once more to diabetes, it is not a kidney disease because of glycosuria, nor a pancreatic disease because of hypoinsulinemia, nor a disease of the pituitary; it is the disease of an organism all of whose functions are changed, which is threatened by tuberculosis, whose supperated infections are endless, whose limbs are rendered useless by arteritis and gangrene; moreover, it can strike man or woman, threaten them with coma, often hit them with impotence or sterility, for whom pregnancy, should it occur, is a catastrophe, whose tears – O irony of secretions! – are sweet.[18] It seems very artificial to break up disease into symptoms or to consider its complications in the abstract. What is a symptom without context or background? What is a complication separated from what it complicates? When an isolated symptom or a functional mechanism is termed pathological, one forgets that what makes them so is their inner relation in the indivisible totality of individual behavior. The situation is such that if the physiological analysis of separated functions is known in the presence of pathological facts, this is due to previous clinical information, for clinical practice puts the physician in contact with complete and concrete individuals and not with organs and their functions. Pathology, whether anatomical or physiological, analyzes in order to know more, but it can be known as pathology, that is, as the study of the mechanisms of disease, only insofar as it receives from clinical practice this notion of disease, whose origin must be sought in the experience men have in their relations with the whole of their environment.

If the above propositions make some sense, how can we then explain that the modern clinician more readily adopts the point of view of the physiologist than that of the sick man? It is undoubtedly because of this massive fact of medical experience, namely that subjective morbid symptoms and objective symptoms rarely overlap. It is simply capricious for a urologist to say that a man who complains of his kidneys is a man who has nothing wrong with his kidneys. For the sick man the kidneys are a cutaneous–muscular territory in the lumbar region, while for the physician they are vital organs connected to others. The well-known fact about reported pains, whose multiple explanations have been very obscure up to now, prevents one from thinking that the pains experienced by the sick man as major subjective symptoms bear a constant relation to the underlying organs to which they seem to call attention. But most of all, the often prolonged latency of certain degeneracies, the inconspicuousness of certain infestations or infections lead the physician to regard the direct pathological experience of the patient as negligible, even to consider it as systematically falsifying the objective pathological fact. Every physician knows, having learned it occasionally to his embarrassment, that the immediate sensible awareness of organic life in itself constitutes neither a science of the same organism nor infallible knowledge of the localization or date of the pathological lesions involving the human body.

Here is perhaps why until now pathology has retained so little of that character which disease has for the sick man – of being really *another way of life*. Certainly pathology is correct in suspecting and rectifying the opinion of the sick man who, because he feels different, thinks he also knows in what and how he is different. It does not follow that because the sick man is clearly mistaken on this second point, he is also mistaken on the first. Perhaps his feeling is the foreshadowing of what contemporary pathology is just beginning to see, namely that the pathological state is not a simple, quantitatively varied extension of the physiological state, but something else entirely.[19*]

Chapter IV

The Conceptions of René Leriche

The invalidity of the sick man's judgment concerning the reality of his own illness is an important theme in a recent theory of disease. This is Leriche's theory, which, though at times rather wavering, is nuanced, concrete and profound. It seems necessary to present and examine it after the preceding theory, which it extends in one direction and from which it clearly deviates in others. "Health," says Leriche, "is life lived in the silence of the organs" [73, *6.16–7*]. Conversely, "disease is what irritates men in the normal course of their lives and work, and above all, what makes them suffer" [73, *6.22–3*]. The state of health is a state of unawareness where the subject and his body are one. Conversely, the awareness of the body consists in a feeling of limits, threats, obstacles to health. Taking these formulae in their full sense, they mean that the actual notion of the normal depends on the possibility of violating the norm. Here at last are definitions which are not empty words, where the relativity of the contrasting terms is correct. For all that the primitive term is not positive; for all that the negative term does not represent nothingness. Health is positive, but not primitive, disease is negative, but in the form of opposition (irritation), not deprivation.

Nevertheless, if neither reservation nor correction is subsequently brought to bear on the definition of health, the definition of disease is immediately straightened out. For this definition of disease is that of the sick man, not that of the doctor; and valuable though it is from the point of view of awareness, it is not the point of view of science. Leriche shows, in effect, that the silence of the organs does not necessarily equal the absence of disease, that there are functional lesions or perturbations which long remain imperceptible to those whose lives they endanger. It is with the frequent delay in feeling our internal irregularities that we pay for the prodigality with which our organism has been constructed, for it has too many of every tissue: more lungs than are strictly required for breathing, more kidneys than are needed to secrete urine to the edge of intoxication. The conclusion is that "if one wants to define disease, it must be dehumanized" [73, *6.22–3*]; and more brutally, "in disease, when all is said and done, the least important thing is man" [73, *6.22–4*]. Hence it is no longer pain or functional incapacity and social infirmity which makes disease, but rather anatomical alteration or physiological disturbance. Disease plays its tricks at the tissue level, and in this sense, there can be sickness without a sick person. Take, for example, a man who has never complained of pathological occurrences and whose life is cut short by murder or a car crash. According to Leriche's theory, if an autopsy of medical–legal intent were to reveal a cancer of the kidney unknown to its late owner, one should conclude in favor of a disease, although there would be no one to whom to attribute it – neither to the cadaver which is no longer competent, nor retroactively to the formerly live man who had no idea of it, having had his life come to an end before the cancer's stage of development at which, in all clinical probability, pain would have finally announced the illness. The disease which never existed in the man's consciousness begins to exist in the physician's science. We think *that there is nothing in science that has not first appeared in the consciousness*, and that in the case now before us, it is particularly the sick man's point of view which forms the basis of truth. And here is why. Doctors and surgeons have clinical information and sometimes use laboratory techniques which allow them to see "patients" in people who do not feel that way. This is a fact. But a fact to be interpreted. It is only because today's practitioners are the heirs to a medical culture transmitted to them by yesterday's practitioners that, in terms of clinical perspicacity, they overtake and outstrip their regular or occasional clients. There has always been a moment when, all things considered, the practitioner's attention has been drawn to certain symptoms, even solely objective ones, by men who were complaining of not being normal – that is, of not being the same as they had been in the past – or of suffering. If, today, the physician's knowledge of disease can anticipate the sick man's experience of it, it is because at one time this experience gave rise to, summoned up, that knowledge. Hence medicine always exists *de jure*, if not *de facto*, because there are men who feel sick, not because there are doctors to tell men of their illnesses. The historical evolution of the relations between the physician and the sick man in clinical consultation changes nothing in the normal, permanent relationship of the sick man and disease.

This critique can be all the more boldly propounded in that Leriche, retracting what was too trenchant in his first formulation, partially confirms it. Carefully distinguishing the static from the dynamic point of view in pathology, Leriche claims complete primacy for the latter. To those who would identify disease and lesion, Leriche objected that the anatomical fact must in reality be considered "second and secondary: second, because it is produced by a primitively functional deviation in the life of the tissues; secondary, because it is only one element in the disease and not the dominant one" [73, *6.76-6*]. Consequently, it is the sick man's disease which very unexpectedly becomes again the adequate concept of disease, more adequate in any case than the concept of the anatomical pathologist.

> The idea must be accepted that the disease of the sick man is not the anatomical disease of the doctor. A stone in an atrophic gall bladder can fail to give symptoms for years and consequently create no disease, although there is a state of pathological anatomy. . . . Under the same anatomical appearances one is sick and one isn't. . . . The difficulty must no longer be conjured away by simply saying that there are silent and masked forms of disease: these are nothing but mere words. The lesion is not enough perhaps to make the clinical disease the disease of the sick man, for this disease is something other than the disease of the anatomical pathologist [*ibid.*].

But it is not a good idea to credit Leriche with more than he has decided to accept. What he in fact means by the sick person is much more the organism in action, in functions, than the individual aware of his organic functions. The sick man in this new definition is not wholly the sick man of the first, the actual man aware of his favored or disfavored situation in life. The sick man has ceased to be an entity for the anatomist but he remains an entity for the physiologist, for Leriche states precisely: "This new representation of disease leads medicine into closer contact with physiology, that is, with the science of functions, and leads it to concern itself at least as much with pathological physiology as with pathological anatomy" [*ibid.*]. Thus, the coincidence of disease and the sick man takes place in the physiologist's science, but not yet in the real man's consciousness. And yet this first coincidence is enough, for Leriche himself provides us with the means to obtain from this the second.

Taking up Claude Bernard's ideas – certainly in full awareness – Leriche also asserts the continuity and indiscernability of the physiological state and the pathological state. For example, in forming the theory of vasoconstrictive phenomena (whose long unrecognized complexity he demonstrated) and their transformation into spasm phenomena, Leriche writes:

> *From tonus to vaso-constriction, that is, to physiological hypertonia, from vaso-constriction to spasm, there is no borderline.* One passes from one state to the other without transition, and it is the effects rather than the thing itself which makes for differentiations. Between physiology and pathology there is no threshold [74, *234*].

Let us understand this last formulation clearly. There is no quantitative threshold which can be detected by objective methods of measurement. But there is nonetheless qualitative distinction and opposition in terms of the different effects of the same quantitatively variable cause.

> Even with perfect conservation of the arterial structure, the spasm, at a distance, has grave pathological effects: it causes pain, produces fragmented or diffuse necroses; last and not least it gives rise to capillary and arterial obliteration at the periphery of the system [74, *234*].

Obliteration, necrosis, pain – these are pathological facts for which physiological equivalents are sought in vain: a blocked artery is, physiologically speaking, no longer an artery, since it is an obstacle, and no longer a path for circulation; physiologically, a necrotic cell is no longer a cell, since, if there is an anatomy of the cadaver, in terms of an etymological definition, there could not exist a physiology of the cadaver; finally, pain is not a physiological sensation because, according to Leriche, "pain is not in nature's plan."

As far as the problem of pain is concerned, Leriche's original and profound thesis is known. It is impossible to consider pain as the expression of a normal activity, of a sense susceptible of permanent exercise, a sense which would exert itself through the organ of specialized, peripheral receptors, of suitable paths of nervous conduction and delimited central analyzers; equally impossible to consider pain either as a detector of and diligent warning signal for events menacing organic integrity from within and without, or as a reaction of salutary defense which the doctor should respect and even reinforce. Pain is "a monstrous individual phenomenon and not a law of the species. A fact of disease" [74, *490*]. We must understand the full importance of these last words. Disease is no longer defined in terms of pain: rather, pain is presented as disease. And what Leriche understands this time as disease is not the quantitative modification of a physiological or normal phenomenon but rather an authentically abnormal state. "Pain-disease in us is like an accident which runs counter to the laws of normal sensation.... Everything about it is abnormal, rebels against the law" [*ibid.*]. At this point Leriche is so sensible of his departure from a classical dogma that he feels the very familiar need to call upon its majesty at the very moment that he is forced to undermine its foundations. "Yes, of course, pathology is never anything but a physiology gone wrong. It was at the Collège de France, in this chair that this idea was born and every day it strikes us as being increasingly true" [74, *482*]. The phenomenon of pain thus verifies electively Leriche's ever-present theory of the state of disease as a "physiological novelty." This conception comes to light in a timid way in the last pages of Vol. VI of the *Encyclopédie française* (1936):

Disease no longer appears to us as a parasite living in and off

> of the man it consumes. We see here the consequence of a deviation – small at first – of the physiological order. In short, it is a new physiological order to which therapeutics must aim to adapt the sick man [73, *6.76–6*].

But this conception is plainly asserted by the following:

> The production of a symptom, even a major one, in a dog, does not mean that we have brought about a human disease. The latter is always an aggregate. That which produces disease in us touches life's ordinary resiliences so subtly that their responses are less that of a physiology gone wrong than that of a new physiology where many things, tuned in a new key, have unusual resonance [76, *11*].

It is not possible for us to examine this theory of pain for its own sake with all the attention it deserves, but we must still indicate its interest for the problem concerning us here. It seems quite important to us that a doctor recognize in pain a phenomenon of total reaction which makes sense, which is a sensation only at the level of concrete human individuality. "Physical pain is not a simple question of nerve impulses moving at a fixed speed along a nerve. *It is the result of the conflict between a stimulant and the individual as a whole*" [74, *488*]. It seems to us quite important that a doctor state that man makes his pain – as he makes a disease or as he makes his mourning – rather than that he receives it or submits to it. Conversely, to consider pain as an impression received at a point of the body and transmitted to the brain is to assume that it is complete in and of itself, without any relation to the activity of the subject who experiences it. It is possible that the inadequacy of anatomical and physiological data in this problem gives Leriche complete freedom, starting from other positive arguments, to deny the specificity of pain. But to deny the anatomic and physiological specificity of a nerve apparatus peculiar to pain is not, in our opinion, necessarily to deny the functional character of pain. Certainly, it is too obvious that pain is not always a faithful and infallible warning signal, that the finalists are kidding themselves by assigning it premonitory capacities and responsibilities which no science of the human body would want to assume. But it is equally obvious that indifference on the part of a living being to his conditions of life, to the quality of his exchanges with his environment, is profoundly abnormal. It can be admitted that pain is a vital sensation without admitting that it has a particular organ or that it has encyclopedic value as a mine of information with regard to the topographical or functional order. The physiologist can indeed denounce the illusions of pain as the physicist does those of sight; this means that sensation is not knowledge and that its normal value is not a theoretical value, but this does not mean that it is normally without value. It seems that one must above all carefully distinguish pain of integumentary [surface] origin from pain of visceral origin. If the latter is presented as abnormal, it seems difficult to dispute the normal character of pain which arises at the surface of the organism's separation from as well as encounter with the environment. The suppression of integumentary pain in scleroderma or syringomyelia can lead to the organism's indifference to attacks on its integrity.

But what we must bear in mind is that Leriche, in defining disease, sees no other way to define it except in terms of its effects. Now with at least one of these effects, pain, we unequivocally leave the plane of abstract science for the sphere of concrete awareness. This time we obtain the total coincidence of disease and the diseased person, for pain-disease, to speak as Leriche does, is a fact at the level of the entire conscious individual, it is a fact which Leriche's fine analyses, relating the participation and collaboration of the whole individual to his pain, allow us to call "behavior."

From here on in we can see clearly in what ways Leriche's ideas extend those of Comte and Bernard and, being subtler and richer in authentic medical experience, in what ways they deviate from them, for with regard to the relations between physiology and pathology Leriche brings to bear the judgment of the technician, not that of the philosopher like Comte or the scientist like Bernard. The idea which Comte and Bernard have in common – despite the difference in intentions mentioned above – is that normally a technology must be the application of a science. This is the fundamental positivist idea: to know in order to act. Physiology must throw light on pathology in order to establish therapeutics. Comte thought that disease served as a substitute for experiments, and Claude Bernard, that experiments, even those performed on animals, led us to the diseases of man. But, in the final analysis, for both men we can progress logically only from experimental physiological knowledge to medical technology. Leriche himself thinks that we progress more often in fact – and should always in theory – from medical and surgical technology prompted by the pathological state to physiological knowledge. Knowledge of the physiological state is obtained by retrospective abstraction from the clinical and therapeutic experience.

> We can ask ourselves whether the study of normal man, even when it is based on that of animals, will ever be enough to inform us fully about the normal life of man. The generosity of the plan on which we are built makes analysis very difficult. Above all, this analysis is carried out by studying the deficiencies produced by the suppression of organs, that is, by introducing variables in the order of life and looking for the conse-

> quences. Unfortunately, with a healthy person experimentation is always a bit brutal in its determinism and the healthy man quickly corrects the slightest spontaneous insufficiency. It is perhaps easier when variables are introduced into man imperceptibly by means of disease, or therapeutically, once disease has struck. The sick man can thus advance knowledge about the normal man. By studying him, deficiencies are discovered in him that the most subtle experiment would fail to produce in animals, and thanks to which normal life can be regained. In this way the complete study of disease tends to become an increasingly essential element of normal physiology [73, *6.76–6*].

Obviously, these ideas are closer to those of Comte than to those of Claude Bernard – but with a big difference. As we have seen, Comte thinks that knowledge of the normal state must normally precede an evaluation of the pathological state and that, strictly speaking, it could be formed – though without the ability to extend very far – without the slightest reference to pathology; similarly, Comte defends the independence of theoretical biology in relation to medicine and therapeutics [27, *247*]. By contrast, Leriche thinks that physiology is the collection of solutions to problems posed by sick men through their illnesses. This is indeed one of the most profound insights on the problem of the pathological: "At every moment there lie within us many more physiological possibilities than physiology would tell us about. But it takes disease to reveal them to us" [76, *11*]. Physiology is the science of the functions and ways of life, but it is life which suggests to the physiologist the ways to explore, for which he codifies the laws. Physiology cannot impose on life just those ways whose mechanism is intelligible to it. Diseases are new ways of life. Without the diseases which incessantly renew the area to be explored, physiology would mark time on well-trod ground. But the foregoing idea can also be understood in another, slightly different sense. Disease reveals normal functions to us at the precise moment when it deprives us of their exercise. Disease is the source of the speculative attention which life attaches to life by means of man. If health is life in the silence of the organs, then, strictly speaking, there is no science of health. Health is organic innocence. It must be lost, like all innocence, so that knowledge may be possible. Physiology is like all science, which, as Aristotle says, proceeds from wonder. But the truly vital wonder is the anguish caused by disease.

It was no exaggeration to announce in the introduction to this chapter that Leriche's conceptions, placed once again in historical perspective, would be able to take on unexpected emphasis. It does not seem possible that any philosophical or medical exploration of the theoretical problems posed by disease can ignore them in the future. At the risk of offending certain minds for whom the intellect is realized only in intellectualism, let me repeat once more that the intrinsic value of Leriche's theory – independent of any criticism applicable to some details of content – lies in the fact that it is the theory of a technology, a theory for which technology exists, not as a docile servant carrying out intangible orders, but as advisor and animator, directing attention to concrete problems and orienting research in the direction of obstacles without presuming anything in advance of the theoretical solutions which will arise.

Chapter V

Implications of a Theory

"Medicine," says Sigerist, "is the most closely linked to the whole of culture, every transformation in medical conceptions being conditioned by transformations in the ideas of the epoch" [107, *42*]. The theory we just expounded, at once medical, scientific and philosophical, perfectly verifies this proposition. It seems to us to satisfy simultaneously several demands and intellectual postulates of the historical moment of the culture in which it was formulated.

First of all there emerges from this theory the conviction of rationalist optimism that evil has no reality. What distinguishes nineteenth-century medicine (particularly before the era of Pasteur) in relation to the medicine of earlier centuries is its resolutely monist character. Eighteenth-century medicine, despite the efforts of the iatromechanists and iatrochemists, and under the influence of the animists and vitalists, remained a dualist medicine, a medical Manichaeanism. Health and Disease fought over man the way Good and Evil fought over the World. It is with a great deal of intellectual satisfaction that we take up the following passage in a history of medicine:

> Paracelsus was a visionary, Van Helmont, a mystic, Stahl, a pietist. All three were innovative geniuses but were influenced

> by their environment and by inherited traditions. What makes appreciation of the reform doctrines of these three great men very hard is the extreme difficulty one experiences in trying to separate their scientific from their religious beliefs.... It is not at all certain that Paracelsus did not believe that he had found the elixir of life; it is certain that Van Helmont identified health with salvation and sickness with sin; and in his account of *Theoria medica vera* Stahl himself, despite his intellectual vigor, availed himself more than he needed to of the belief in original sin and the fall of man [48, *311*].

More than he needed to! says the author, quite the great admirer of Broussais, sworn enemy at the dawn of the nineteenth century of all medical ontology. The denial of an ontological conception of disease, a negative corollary of the assertion of a quantitative identity between the normal and the pathological, is first, perhaps, the deeper refusal to confirm evil. It certainly cannot be denied that a scientific therapeutics is superior to a magical or mystical one. It is certain that knowledge is better than ignorance when action is required, and in this sense the value of the philosophy of the Enlightenment and of positivism, even scientistic, is indisputable. It would not be a question of exempting doctors from the study of physiology and pharmacology. It is very important not to identify disease with either sin or the devil. But it does not follow from the fact that evil is not a being, that it is a concept devoid of meaning; it does not follow that there are no negative values, even among vital values; it does not follow that the pathological state is essentially nothing other than the normal state.

Conversely, the theory in question conveys the humanist conviction that man's action on his environment and on himself can and must become completely one with his knowledge of the environment and man; it must be normally only the application of a previously instituted science. Looking at the *Leçons sur le diabète* [Lectures on Diabetes] it is obvious that if one asserts the real homogeneity and continuity of the normal and the pathological it is in order to establish a physiological science that would govern therapeutic activity by means of the intermediary of pathology. Here the fact that human consciousness experiences occasions of new growth and theoretical progress in its domain of nontheoretical, pragmatic and technical activity is not appreciated. To deny technology a value all its own outside of the knowledge it succeeds in incorporating, is to render unintelligible the irregular way of the progress of knowledge and to miss that overtaking of science by the power which the positivists have so often stated while they deplored it. If technology's rashness, unmindful of the obstacles to be encountered, did not constantly anticipate the prudence of codified knowledge, the number of scientific problems to resolve, which are surprises after having been setbacks, would be far fewer. Here is the truth that remains in empiricism, the philosophy of intellectual adventure, which an experimental method, rather too tempted, by reaction, to rationalize itself, failed to recognize.

Nevertheless, Claude Bernard cannot be reproached – without our being inaccurate – for having ignored the intellectual stimulus found by physiology in clinical practice. He himself acknowledged the fact that his experiments on glycemia and glucose production in the animal organism have as their point of departure observations related to diabetes and the disproportion sometimes noticeable between the amount of carbohydrates ingested and the amount of glucose eliminated by the urine. He himself formulated the following general principal: "The medical problem must first be posed so that it is given by observation of the disease, and then the pathological phenomena must be analyzed experimentally as one tries to provide a physiological explanation for them" [6, *349*]. Despite everything, it is still true that for Bernard the pathological fact and its physiological explanation do not have the same theoretical importance. The pathological fact accepts explanation more than it stimulates it. This is even more obvious in the following text: "Diseases are essentially nothing but physiological phenomena in new conditions which have to be determined" [6, *346*]. For whoever knows physiology, diseases verify the physiology he knows, but essentially they teach him nothing; phenomena are the same in the pathological state, save for conditions. As if one could determine a phenomenon's essence apart from its conditions! As if conditions were a mask or frame which changed neither the face nor the picture! One should compare this proposition with that of Leriche cited above in order to feel all the expressive importance of a verbal nuance: "At every moment there lie within us many more physiological possibilities than physiology tells us about. But it takes disease to reveal them to us."

Here again we owe to the chance of bibliographical research the intellectual pleasure of stating once more that the most apparently paradoxical theses also have their tradition which undoubtedly expresses their permanent logical necessity. Just when Broussais was lending his authority to the theory which established physiological medicine, this same theory was provoking the objections of an obscure physician, one Dr. Victor Prus, who was rewarded by the Société de Médecine du Gard in 1821 for a report entered in a competition whose object was the precise definition of the terms phlegmasia and irritation and their importance for practical medicine. After having challenged the idea that physiology by itself forms the natural foundation of medicine; that it alone can ever establish the knowledge of symptoms, their relationships and their value; that pathological anatomy can ever be deduced from the knowledge of normal phenomena; that the prognosis of diseases derives from the knowledge of physiological laws, the author adds:

> If we want to exhaust the question dealt with in this article we would have to show that *physiology, far from being the foundation of pathology, could only arise in opposition to it*. It is through the changes which the disease of an organ and sometimes the complete suspension of its activity transmit to its functions that we learn the organ's use and importance.... Hence an exostosis, by compressing and paralyzing the optic nerve, the brachial nerves, and the spinal cord, shows us their usual destination. Broussonnet lost his memory of substantive words; at his death an abscess was found in the anterior part of his brain and one was led to believe that that is the center for the memory of names.... Thus pathology, aided by pathological anatomy, has created physiology: every day pathology clears up physiology's former errors and aids its progress [95, *L*].

In writing the *Introduction à l'étude de la médecine expérimentale*, Claude Bernard set out to assert not only that efficacious action is the same as science, but also, and analogously, that science is identical with the discovery of the laws of phenomena. On this point his agreement with Comte is total. What Comte in his philosophical biology calls the doctrine of the conditions of existence, Bernard calls determinism. He flatters himself with having been the first to introduce that term into scientific French.

> I believe I am the first to have introduced this word to science, but it has been used by philosophers in another sense. It will be useful to determine the meaning of this word in a book which I plan to write: *Du déterminisme dans les sciences* [On Determinism in the Sciences]. This will amount to a second edition of my *Introduction à la médecine expérimentale* [103, 96].

It is faith in the universal validity of the determinist postulate which is asserted by the principal "physiology and pathology are one and the same thing." At the very time that pathology was saddled with prescientific concepts, a physicochemical physiology existed which met the demands of scientific knowledge, that is, a physiology of quantitative laws verified by experimentation. Understandably, early nineteenth-century physicians, justifiably eager for an effective, rational pathology, saw in physiology the prospective model which came closest to their ideal.

> Science rejects the *indeterminate*, and in medicine, when opinions are based on medical palpation, inspiration, or a more or less vague intuition about things, we are outside of science and are given the example of this medicine of fantasy, capable of presenting the gravest perils as it delivers the health and lives of sick men to the whims of an inspired ignoramus [6, *96*].

But just because, of the two – physiology and pathology – only the first involved laws and postulated the determinism of its object, it was not necessary to conclude that, given the legitimate desire for a rational pathology, the laws and determinism of pathological facts are the same laws and determinism of physiological facts. We know the antecedents of this point of doctrine from Bernard himself. In the lecture devoted to the life and works of Magendie at the beginning of the *Leçons sur les substances toxiques et médicamenteuses* [Lectures on Toxic and Medicinal Substances] (1857), Bernard tells us that the teacher whose chair he occupies and whose teaching he continues "drew the feeling of real science" from the illustrious Laplace. We know that Laplace had been Lavoisier's collaborator in the research on animal respiration and animal heat, the first brilliant success in research on the laws of biological phenomena following the experimental and measuring methods endorsed by physics and chemistry. As a result of this work Laplace had retained a distinct taste for physiology and he supported Magendie. If Laplace never used the term "determinism," he is one of its spiritual fathers and, at least in France, an authoritative and authorized father of the doctrine designated by the term. For Laplace determinism is not a methodological requirement, a normative research postulate sufficiently flexible to prejudice in any way the form of the results to which it leads: it is reality itself, complete, cast *ne varietur* in the framework of Newtonian and Laplacian mechanics. Determinism can be conceived as being *open* to incessant corrections of the formulae of laws and the concepts they link together, or as being *closed* on its own assumed definitive content. Laplace constructed the theory of closed determinism. Claude Bernard did not conceive of it in any other way and this is undoubtedly why he did not believe that the collaboration of pathology and physiology could lead to a progressive rectification of physiological concepts. It is appropriate here to recall Whitehead's dictum:

> Every special science has to assume results from other sciences. For example, biology presupposes physics. It will usually be the case that these loans really belong to the state of science thirty or forty years earlier. The presuppositions of the physics of my boyhood are today powerful influences in the mentality of physiologists.[20]

Finally, as a result of the determinist postulate, it is the reduction of quality to quantity which is implied by the essential identity of physiology and pathology. To reduce the difference between a healthy man and a diabetic to a quantitative difference of the amount of glucose within the body; to delegate the task of distinguishing one who is diabetic from one who is not to a renal threshold conceived simply as a quantitative difference of level, means obeying the spirit of the physical sciences which, in buttressing phenomena with laws, can explain them only in terms of their reduction to a common measure. In order to introduce terms into the relationships of composition and dependence, the homogeneity of these terms should be obtained first. As Emile Meyerson has shown, the human spirit attained knowledge by identifying reality and quantity. But it should be remembered that, though scientific knowledge invalidates qualities, which it makes appear illusory, for all that it does not annul them. Quantity is quality denied, but not quality suppressed. The qualitative variety of simple lights, perceived as colors by the human eye, is reduced by science to the quantitative difference of wavelengths, but the qualitative variety still persists in the form of quantitative differences in the calculation of wavelengths. Hegel maintains that, by its growth or diminution, quantity changes into quality. This would be perfectly inconceivable if a relation to quality did not still persist in the negated quality which is called quantity.[21]

From this point of view it is completely illegitimate to maintain that the pathological state is really and simply a greater or lesser variation of the physiological state. Either this physiological state is conceived as having one quality and value for the living man, and so it is absurd to extend that value, identical to itself in its variations, to a state called pathological whose value and quantity are to be differentiated from and essentially contrasted with the first. – Or what is understood as the physiological state is a simple summary of quantities, without biological value, a simple fact or system of physical and chemical facts, but as this state has no vital quality, it cannot be called healthy or normal or physiological. Normal and pathological have no meaning on a scale where the biological object is reduced to colloidal equilibria and ionized solutions. In studying a state which he describes as physiological, the physiologist qualifies it as such, even unconsciously; he considers this state as positively qualified by and for the living being. Now this qualified physiological state is not, as such, what is extended, identically to itself, to another state capable of assuming, inexplicably, the quality of morbidity.

Of course this is not to say that an analysis of the conditions or products of pathological functions will not give the chemist or physiologist numerical results comparable to those obtained in a way consistent with the terms of the same analyses concerning the corresponding, so-called physiological functions. But it is arguable as to whether the terms *more* and *less*, once they enter the definition of the pathological as a quantitative variation of the normal, have a purely quantitative meaning. Also arguable is the logical coherence of Bernard's principal: "The disturbance of a normal mechanism, consisting in a quantitative variation, an exaggeration, or an attenuation, constitutes the pathological state." As has been pointed out in connection with Broussais's ideas, in the order of physiological functions and needs, one speaks of more and less in relation to a norm. For example, the hydration of tissues is a fact which can be expressed in terms of more and less; so is the percentage of calcium in blood. These quantitatively different results would have no quality, no value in a laboratory, if the laboratory had no relationship with a hospital or clinic where the results take on the value or not of uremia, the value or not of tetanus. Because physiology stands at the crossroads of the laboratory and the clinic, two points of view about biological phenomena are adopted there, but this does not mean that they can be interchanged. The substitution of quantitative progression for qualitative contrast in no way annuls this opposition. It always remains at the back of the mind of those who have chosen to adopt the theoretical and metrical point of view. When we say that health and disease are linked by all the intermediaries, and when this continuity is converted into homogeneity, we forget that the difference continues to manifest itself at the extreme, without which the intermediaries could in no way play their mediating role; no doubt unconsciously, but wrongly, we confuse the abstract calculation of identities and the concrete appreciation of differences.

Part Two

Do Sciences of the Normal and the Pathological Exist?

Chapter I

Introduction to the Problem

It is interesting to note that in their own discipline contemporary psychiatrists have brought about a rectification and restatement of the concepts of *normal* and *pathological* from which physicians and physiologists apparently have not cared to draw a lesson concerning themselves. Perhaps the reason for this is to be sought in the usually closer relations between psychiatry and philosophy through the intermediary of psychology. In France, Blondel, Daniel Lagache and Eugène Minkowski in particular have contributed to a definition of the general essence of the morbid or abnormal psychic fact and its relations with the normal. In his *La conscience morbide* [Morbid Consciousness (Paris, Alcan, 1914)], Blondel describes cases of insanity where the patients seem incomprehensible to others as well as to themselves, where the doctor really has the impression of dealing with another mental structure; he seeks the explanation for this in the impossible situation where these patients translate the data of their cenesthesia into the concepts of normal language. It is impossible for the physician, starting from the accounts of sick men, to understand the experience lived by the sick man, for what sick men express in ordinary concepts is not directly their experience but their interpretation of an experience for which they have been deprived of adequate concepts.

Lagache is quite far from this pessimism. He thinks that a distinction must be made in the abnormal consciousness between variations of nature and variations of degree; in certain psychoses the patient's personality is heterogenous with the former personality, in others, one is the extension of the other. Along with Jaspers, Lagache distinguishes incomprehensible psychoses from comprehensible ones; in the latter case the psychosis seems to be intelligibly related to the earlier psychic life. Hence, aside from difficulties posed by the general problem of understanding others, psychopathology is a source of documents which can be utilized in general psychology, a source of light to be shed on normal consciousness [66, *8.08–8*]. But – and this is the point we want to make – this position is quite different from Ribot's mentioned above. Disease, according to Ribot, is a spontaneous and methodological substitute for experimentation, reaches the unreachable, but respects the nature of the normal elements to which it reduces psychic functions. Disease disorganizes but does not transform, it reveals without altering. Lagache does not admit the assimilation of disease with experimentation. Experimentation demands an exhaustive analysis of the phenomenon's conditions of existence and a rigorous determination of the conditions which are made to vary in order to observe the repercussions. On none of these points is mental illness comparable with experimentation. First, "nothing is less well known than the conditions in which nature establishes these experiences, these mental illnesses: the beginning of a psychosis most often escapes the notice of the doctor, the patient, and those surrounding him; its physiopathology, its pathological anatomy are obscure" [66, *8.08–5*]. Later: "at the basis of the illusion which assimilates the pathological method in psychology with the experimental method, there is the atomistic and associationist representation of mental life; this is the faculty psychology" [*ibid.*]. As there are no separable elementary psychic facts, pathological symptoms cannot be compared with elements of normal consciousness because a symptom has a pathological significance only in its clinical context, which expresses a global disturbance. For example, a verbal psychomotor hallucination is involved in delirium and delirium is involved in an alteration of the personality [66, *8.08–7*]. Consequently, general psychology can use psychopathological data in the same epistemologically valid way as facts observed in normal people, but not without one express adaptation for the originality of the pathological. Unlike Ribot, Lagache thinks that morbid disorganization is not the symmetrical inverse of normal organization. Forms can exist in pathological consciousness which have no equivalent in the normal state and yet by which general psychology is enriched:

> Even the most heterogeneous structures, beyond the intrinsic interest of their study, can furnish data for problems posed by general psychology; they even pose new problems, and a curious peculiarity of psychopathological vocabulary is its accommodation of negative expressions without equivalent in normal psychology; how can we fail to recognize the new light thrown on our knowledge of the human being by ideas such as that of discordance? [66, *8.08–8*].

Minkowski also thinks that the fact of insanity cannot be reduced to just the one fact of disease, determined by its reference to one image or precise idea of the average or normal being. When we call another man insane, we do so intuitively "as men, not as specialists." The madman is "out of his mind" not so much in relation to other men as to life: he is not so much deviant as different. "Through anomalies a human being detaches himself from everything which forms men and life. In a particularly radical and striking – and therefore primitive – way they reveal to us the significance of an altogether 'singular' form of being. This circumstance explains why 'being sick' does not at all exhaust the phenomenon of insanity, which, coming to our attention from the perspective of 'being different' in the qualitative sense of the word, directly opens the way to psychopathological considerations made from that perspective" [84, 77]. According to Minkowski, insanity or a psychic anomaly presents its own features which he believes are not contained in the concept of disease. First of all in an anomaly there is the primacy of the negative; evil is detached from life while good is enmeshed with vital dynamism and finds its meaning only "in a constant progression called to extend every conceptual formula relative to this would-be norm" [84, *78*]. Isn't it the same in the realm of the body and there too doesn't one speak of health only because diseases exist? But according to Minkowski mental illness is a more immediately vital category than disease: somatic disease is capable of a superior empirical precision, of a better-defined standardization; somatic disease does not rupture the harmony between fellow creatures, the sick man is for us what he is for himself, whereas the psychically abnormal has no consciousness of his state. "The individual dominates the sphere of mental deviations much more than he does in the somatic sphere" [84, *79*].

We do not share Minkowski's opinion on this last point. Like Leriche we think that health is life in the silence of the organs, that consequently the biologically normal, as we have already said, is revealed only through infractions of the norm and that concrete or scientific awareness of life exists only through disease. We agree with Sigerist that "disease isolates" [107, *86*], and that even if "this isolation does not alienate men but on the contrary brings them closer to the sick man" [107, *95*], no perceptive patient can ignore the renunciations and limitations imposed by healthy men in order to come near him. We agree with Goldstein that the norm in pathology is above all an individual norm [46, *272*]. In short, we think that to consider life as a dynamic force of transcendence as Minkowski does (whose sympathies for Bergsonian philosophy are revealed in works such as *La schizophrénie* [Paris, Payot, 1927] or *Le temps vécu* [Neuchâtel, Delachaux and Niestlé, 1968; translated as *Lived Time*, Evanston, Northwestern University Press, 1970]) is to force oneself to treat somatic anomaly and psychic anomaly in the same way. When Ey, who approves Minkowski's views, states:

> The normal man is not a mean correlative to a social concept, it is not a judgment of reality but rather a judgment of value; it is a limiting notion which defines a being's maximum psychic capacity. There is no upper limit to normality [84, *93*],

we find it sufficient to replace "psychic" with "physical" in order to obtain a very correct definition of the concept of the normal which the physiology and medicine of organic diseases use every day without caring enough to state its meaning precisely.

Moreover, this insouciance has good reasons behind it, particularly on the part of the practicing physician. In the final analysis it is the patients who most often decide – and from very different points of view – whether they are no longer normal or whether they have returned to normality. For a man whose future is almost always imagined starting from past experience, becoming normal again means taking up an interrupted activity or at least an activity deemed equivalent by individual tastes or the social values of the milieu. Even if this activity is reduced, even if the possible behaviors are less varied, less supple than before, the individual is not always so particular as all that. The essential thing is to be raised from an abyss of impotence or suffering where the sick man *almost died*; the essential thing is *to have had a narrow escape*. Take, for example, a young man examined recently, who fell on a moving circular saw, whose arm was deeply cut cross-wise three-fourths the way up but where the internal vascular nerve bundle was unharmed. A quick and intelligent operation allowed the arm to be saved. The arm shows an atrophy of all the muscles, including the forearm. The whole limb is cold, the hand is cyanotic. When stimulated electrically, the group of extensor muscles shows a distinctly degenerated reaction. The movements of flexion, extension and supination of the forearm are limited (flexion limited to 45°, extension to about 170°); pronation is nearly normal. The patient is happy to know that there is the possibility he will recover much of the use of his limb. Certainly, with respect to the other arm, the injured and surgically restored arm will not be normal from the trophic and functional point of view. But *on the whole* the man will take up the trade again which he had chosen or which circumstances put forward, if not imposed; on which, in any case, he places a reason – even a mediocre one – for living. From now on, even if this man obtains equivalent technical results using different procedures of complex gesticulation, socially he will continue to be appreciated according to former norms; he will always be a cartwright or a driver and not a former cartwright or a former driver. The sick man loses sight of the fact that because of his injury he will from now on lack a wide range of neuromuscular adaptations and improvisations, that is, the capacity which perhaps he had never used to better his output and surpass himself, but then only because of lack of opportunity. The sick man maintains that he is not in any *obvious* sense disabled. This notion of *disability* should be studied by a medical expert who would not see in the organism merely a machine whose output must be calculated, an expert who is enough of a psychologist to appreciate lesions as deteriorations more than as percentages.[22*] But in general the experts practice psychology only in order to track down psychoses of reclaiming rights [*psychoses de revendication*] in the subjects presented to them and to talk of pithiatism [morbidity curable by suggestion]. Be that as it may, the practicing physician is very often happy to agree with his patients in defining the normal and abnormal according to their individual norms, except, of course, in the case of gross ignorance on their part of the minimal anatomical and physiological conditions of plant or animal life. We remember having seen in surgical service a simple-minded farmhand both of whose tibias had been fractured by a cart wheel and whom his master had not had treated for fear of who knows what responsibilities; the tibias had joined together by themselves at an obtuse angle. The man had been sent to the hospital after the denunciation by neighbors. It was necessary to rebreak his tibias and set them properly. It is clear that the head of the department who made the decision had another image of the human leg than that of that poor devil and his master. It is also clear that he adopted a norm which would not have satisfied either a Jean Bouin [French Olympic runner in 1912] or a Serge Lifar [dancer, choreographer and ballet master, Paris Opera Ballet, 1930–1958].

Jaspers saw clearly what difficulties lie in this medical determination of the normal and health:

> It is the physician who searches the least for the meaning of the words "health and disease." He is concerned with vital phenomena from the scientific point of view. More than the physicians' judgment, it is the patients' appraisal and the dominant ideas of the social context, which determine what is called "disease" [59, *5*].

What one finds in common in the different meanings given today or in the past to the concept of disease is that they form a judgment of virtual value. "Disease is a general concept of non-value which includes all possible negative values" [59, *9*]. To be sick is to be harmful or undesirable or socially devalued, etc. On the other hand, from the physiological point of view what is desired in health is obvious and this gives the concept of physical disease a relatively stable meaning. Desirable values are "life, a long life, the capacity for reproduction and for physical work, strength, resistance to fatigue, the absence of pain, a state in which one notices the body as little as possible outside of the joyous sense of existence" [59, 6]. However, medical science does not consist in speculating about these common concepts in order to obtain a general concept of disease; its real task is to determine what are the vital phenomena with regard to which men call themselves sick, what are their origins, their laws of evolution, the actions which modify them. The general concept of value is specified in a multitude of concepts of existence. But despite the apparent disappearance of any value judgment in these empirical concepts, the physician persists in talking of diseases, because medical activity, through clinical questioning and therapeutics, has a relationship with the patient and his value judgments [59, 6].

It is perfectly understandable, then, that physicians are not interested in a concept which seems to them to be too vulgar or too metaphysical. What interests them is diagnosis and cure. In principle, curing means restoring a function or an organism to the norm from which they have deviated. The physician usually takes the norm from his knowledge of physiology – called the science of the normal man – from his actual experience of organic functions, and from the common representation of the norm in a social milieu at a given moment. Of the three authorities, physiology carries him furthest. Modern physiology is presented as a canonical collection of functional constants related to the hormonal and nervous functions of regulation. These constants are termed normal insofar as they designate average characteristics, which are most frequently practically observable. But they are also termed normal because they enter ideally into that normative activity called therapeutics. Physiological constants are thus normal in the statistical sense, which is a descriptive sense, and in the therapeutic sense, which is a normative sense. But the question is whether it is medicine which converts – and how? – descriptive and purely theoretical concepts into biological ideals or whether medicine, in admitting the notion of facts and constant functional coefficients from physiology would not also admit – probably unbeknownst to the physiologists – the notion of norm in the normative sense of the word. And it is a question of whether medicine, in doing this, wouldn't take back from physiology what it itself had given. This is the difficult problem to examine now.

CHAPTER II

A Critical Examination of Certain Concepts: The Normal, Anomaly and Disease; The Normal and the Experimental

Littré and Robin's *Dictionnaire de médecine* defines the normal as follows: normal (*normalis*, from *norma*, rule): that which conforms to the rule, regular. The brevity of this entry in a medical dictionary does not surprise us given the observations we have just made. Lalande's *Vocabulaire technique et critique de la philosophie* is more explicit. Since *norma*, etymologically, means a T-square, normal is that which bends neither to the right nor left, hence that which remains in a happy medium; from which two meanings are derived: (1) normal is that which is such that it ought to be; (2) normal, in the most usual sense of the word, is that which is met with in the majority of cases of a determined kind, or that which constitutes either the average or standard of a measurable characteristic. In the discussion of these meanings it has been pointed out how ambiguous this term is since it designates at once a fact and "a value attributed to this fact by the person speaking, by virtue of an evaluative judgment for which he takes responsibility." One should also stress how this ambiguity is deepened by the realist philosophical tradition which holds that, as every generality is the sign of an essence, and every perfection the realization of the essence, a generality observable in fact takes the value of realized perfection, and a common characteristic, the value of an ideal type.

Finally, an analogous confusion in medicine should be emphasized, where the normal state designates both the habitual state of the organs, and their ideal, since the reestablishment of this habitual ideal is the ordinary aim of therapeutics [67].

It seems to us that this last remark has not been developed as it should be and that, in particular, in the entry cited, not enough has been deduced from it concerning the ambiguity of meaning in the term *normal* where one is happy to point out its existence rather than see in it a problem to solve. It is true that in medicine the normal state of the human body is the state one wants to reestablish. But is it because therapeutics aims at this state as a good goal to obtain that it is called normal, or is it because the interested party, that is, the sick man, considers it normal that therapeutics aim at it? We hold the second statement to be true. We think that medicine exists as the art of life because the living human being himself calls certain dreaded states or behaviors pathological (hence requiring avoidance or correction) relative to the dynamic polarity of life, in the form of a negative value. We think that in doing this the living human being, in a more or less lucid way, extends a spontaneous effort, peculiar to life, to struggle against that which obstructs its preservation and development taken as norms. The entry in the *Vocabulaire philosophique* seems to assume that value can be attributed to a biological fact only by "him who speaks," obviously a man. We, on the other hand, think that the fact that a living man reacts to a lesion, infection, functional anarchy by means of a disease, expresses the fundamental fact that life is not indifferent to the conditions in which it is possible, that life is polarity and thereby even an unconscious position of value; in short, life is in fact a normative activity. *Normative*, in philosophy, means every judgment which evaluates or qualifies a fact in relation to a norm, but this mode of judgment is essentially subordinate to that which establishes norms. Normative, in the fullest sense of the word, is that which establishes norms. And it is in this sense that we plan to talk about biological normativity. We think that we are as careful as anyone as far as the tendency to fall into anthropomorphism is concerned. We do not ascribe a human content to vital norms but we do ask ourselves how normativity essential to human consciousness would be explained if it did not in some way exist in embryo in life. We ask ourselves how a human need for therapeutics would have engendered a medicine which is increasingly clairvoyant with regard to the conditions of disease if life's struggle against the innumerable dangers threatening it were not a permanent and essential vital need. From the sociological point of view it can be shown that therapeutics was first a religious, magical activity, but this does not negate the fact that therapeutic need is a vital need, which, even in lower living organisms (with respect to vertebrate structure) arouses reactions of hedonic value or self-healing or self-restoring behaviors.

The dynamic polarity of life and the normativity it expresses account for an epistemological fact of whose important significance Bichat was fully aware. Biological pathology exists but there is no physical or chemical or mechanical pathology:

> There are two things in the phenomena of life: (1) the state of health; (2) the state of disease, and from these two distinct sciences derive: physiology, which concerns itself with the phenomena of the first state, pathology, with those of the second. The history of phenomena in which vital forces have their natural form leads us, consequently, to the history of phenomena where these forces are changed. Now, in the physical sciences only the first history exists, never the second. Physiology is to the movement of living bodies what astronomy, dynamics, hydraulics, hydrostatics, etc. are to inert ones: these last have no science at all which corresponds to them as pathology corres-

> ponds to the first. For the same reason the whole idea of medication is distasteful to the physical sciences. Any medication aims at restoring certain properties to their natural type: as physical properties never lose this type, they do not need to be restored to it. Nothing in the physical sciences corresponds to what is therapeutics in the physiological sciences [13, *I*, *20–21*].

It is clear from this text that natural type must be taken in the sense of normal type. For Bichat the natural is not the effect of a determinism, but the term of a finality. And we know well everything that can be found wrong in such a text from the point of view of a mechanist or materialist biology. One might say that long ago Aristotle believed in a pathological mechanics since he admitted two kinds of movements: natural movements through which a body regains its proper place where it thrives at rest, as a stone goes down to the ground, and fire, up to the sky; – and violent movements by which a body is pushed from its proper place, as when a stone is thrown in the air. It can be said that with Galileo and Descartes, progress in knowledge of the physical world consisted in considering all movements as natural, that is, as conforming to the laws of nature, and that likewise progress in biological knowledge consisted in unifying the laws of natural life and pathological life. It is precisely this unification which Comte dreamed of and Claude Bernard flattered himself with having accomplished, as was seen above. To the reservations which we felt obliged to set forth at that time, let us add this. In establishing the science of movement on the principle of inertia, modern mechanics in effect made the distinction between natural and violent movements absurd, as inertia is precisely an indifference with respect to directions and variations in movement. Life is far removed from such an indifference to the conditions which are made for it; life is polarity. The simplest biological nutritive system of assimilation and excretion expresses a polarity. When the wastes of digestion are no longer excreted by the organism and congest or poison the internal environment, this is all indeed according to law (physical, chemical, etc.) but none of this follows the norm, which is the activity of the organism itself. This is the simple fact that we want to point out when we speak of biological normativity.

There are some thinkers whose horror of finalism leads them to reject even the Darwinian idea of selection by the environment and struggle for existence because of both the term selection, obviously of human and technological import, and the idea of advantage which comes into the explanation of the mechanism of natural selection. They point out that most living beings are killed by the environment long before the inequalities which they can produce even have a chance to be of use to them because it kills above all sprouts, embryos or the young. But as Georges Teissier observed, the fact that many organisms die before their inequalities serve them does not mean that the presentation of inequalities is biologically indifferent [111]. This is precisely the one fact we ask to be granted. There is no biological indifference, and consequently we can speak of biological normativity. There are healthy biological norms and there are pathological norms, and the second are not the same as the first.

We did not refer to the theory of natural selection unintentionally. We want to draw attention to the fact that what is true of the expression *natural selection* is also true of the old expression *vis medicatrix naturae*. Selection and medicine are biological techniques practiced deliberately and more or less rationally by man. When we speak of natural selection or natural medicinal activity we are victims of what Bergson calls the illusion of retroactivity if we imagine that vital prehuman activity pursues goals and utilizes means comparable to those of men. But it is one thing to think that natural selection would utilize anything that resembles *pedigrees*, and *vis medicatrix*, cupping glasses, and another to think that human technique extends vital impulses, at whose service it tries to place systematic knowledge which would deliver them from much of life's costly trial and error.

The expressions "natural selection" and "natural medicinal activity" have one drawback in that they seem to set vital techniques within the framework of human techniques when it is the opposite which seems true. All human technique, including that of life, is set within life, that is, within an activity of information and assimilation of material. It is not because human technique is normative that vital technique is judged such by comparison. Because life is activity of information and assimilation it is the root of all technical activity. In short, we speak of natural medicine in quite a retroactive and, in one sense, mistaken way, but even if we were to assume that we have no right to speak of it, we are still free to think that no living being would have ever developed medical technique if the life within him – as within every living thing – were indifferent to the conditions it met with, if life were not a form of reactivity polarized to the variations of the environment in which it develops. This was seen very well by Guyénot:

> It is a fact that the organism has an aggregate of properties which belong to it alone, thanks to which it withstands multiple destructive forces. Without these defensive reactions, life would be rapidly extinguished.... The living being is able to find instantaneously the reaction which is useful vis-à-vis substances with which neither it nor its kind has ever had contact. The organism is an incomparable chemist. It is the first among physicians. The fluctuations of the environment are almost always a menace to its existence. The living being could not survive if it did not possess certain essential properties. Every injury

would be fatal if tissues were incapable of forming scars and blood incapable of clotting [52, *186*].

By way of summary, we think it very instructive to consider the meaning that the word "normal" assumes in medicine, and the fact that the concept's ambiguity, pointed out by Lalande, is greatly clarified by this, with a quite general significance for the problem of the normal. It is life itself and not medical judgment which makes the biological normal a concept of value and not a concept of statistical reality. For the physician, life is not an object but rather a polarized activity, whose spontaneous effort of defense and struggle against all that is of negative value is extended by medicine by bringing to bear the relative but indispensable light of human science.

Lalande's *Vocabulaire philosophique* contains an important remark about the terms *anomaly* and *abnormal*. *Anomaly* is a substantive with no corresponding adjective at present; *abnormal*, on the other hand, is an adjective with no substantive, so that [French] usage has coupled them, making abnormal the adjective of anomaly. It is quite true that "anomalous" [*anomal*], which Isidore Geoffroy Saint-Hilaire was still using in 1836 in his *Histoire des anomalies de l'organisation* and which also appears in Littré and Robin's *Dictionnaire de médecine*, has fallen into disuse. Lalande's *Vocabulaire* shows that confusion of an etymological nature has helped draw anomaly and abnormal closer together. "Anomaly" comes from the Greek *anomalia* which means unevenness, asperity; *omalos* in Greek means that which is level, even, smooth, hence "anomaly" is, etymologically, *an-omalos*, that which is uneven, rough, irregular, in the sense given these words when speaking of a terrain.[23] A mistake is often made with the etymology of "anomaly," by deriving it not from *omalos* but from *nomos* which means law, hence the compound *a-nomos*. This etymological error is found right in Littré and Robin's *Dictionnaire de médecine*. The Greek *nomos* and the Latin *norma* have closely related meanings, law and rule tending to become confused. Hence, in a strictly semantic sense "anomaly" points to a fact, and is a descriptive term, while "abnormal" implies reference to a value and is an evaluative, normative term; but the switching of good grammatical methods has meant a confusion of the respective meanings of anomaly and abnormal. "Abnormal" has become a descriptive concept and "anomaly," a normative one. Geoffroy Saint-Hilaire, who makes the etymological error, repeated after him by Littré and Robin, tries to maintain the purely descriptive and theoretical meaning of "anomaly," which is a biological fact and must be treated as such, that is, it must be explained, not evaluated, by natural science:

> The word *anomaly,* like the word *irregularity*, must never be taken in the sense which would be deduced literally from its etymological composition. There are no organic formations which are not subject to laws; and the word *disorder*, taken in its real sense, would not be applicable to any productions of nature. "Anomaly" is an expression which has been recently introduced into anatomical language, whose use there is even infrequent. On the other hand, the zoologists from whom it was borrowed, use it very often; they apply it to a large number of animals, who, because of their *unusual* organization and features, find themselves isolated, so to speak, in the series and have only very distant kinship with others in the same class [43, *I, 96, 37*].

According to Geoffroy Saint-Hilaire, it is wrong to speak of either peculiarities of nature, or disorder or irregularity with regard to such animals. If there is an exception, it is to the laws of naturalists, not to the laws of nature, for in nature all species *are what they must be*, equally presenting variety in unity and unity in variety [43, *I, 37*]. In anatomy the term "anomaly" must strictly maintain its meaning of *unusual*, *unaccustomed*; to be *anomalous* is to be removed, in terms of one's organization, from the vast majority of beings to which one must be compared [*ibid.*].

Having defined anomaly in general from the morphological point of view, Geoffroy Saint-Hilaire relates it directly to two biological facts, *the specific type* and *individual variation*. On the one hand, all living species present for examination a multitude of variations in the form and proportional volume of organs; on the other hand, there is a complex of traits "common to the vast majority of individuals who compose a species" and this complex defines the specific type. "Every deviation of the specific type, or in other words, every organic particularity introduced by an individual when compared with the vast majority of the individuals of his species, age, and sex, constitutes what can be called an Anomaly" [43, *I, 30*]. It is clear that, so defined, anomaly is, generally speaking, a purely empirical or descriptive concept, a statistical deviation.

One problem which immediately presents itself is whether the concepts anomaly and monstrosity must be considered equivalent. Geoffroy Saint-Hilaire is on the side of distinction: monstrosity is one species of the genus anomaly. Whence the division of anomalies into *Varieties*, *Structural defects*, *Heterotaxy* and *Monstrosities*. *Varieties* are simple, slight anomalies which do not obstruct the performance of any function and produce no deformity; for example: a supernumerary muscle, and double renal artery. *Structural defects* are simple anomalies, slight in terms of the anatomical relationship, but they make the performance of one or more functions impossible or produce a deformity; for example, a defective anus, hypospadias or harelip. *Heterotaxies*, a term created by

Geoffroy Saint-Hilaire, are complex anomalies, serious in appearance in terms of the anatomical relationship, but they impede no function and are not apparent on the outside; the most remarkable, though rare, example, according to Geoffroy Saint-Hilaire, is the complete transposition of the viscera or *situs inversus*. We know that, while rare, the heart on the right-hand side is no myth. Finally, *Monstrosities* are very complex anomalies, very serious, making the performance of one or more functions impossible or difficult, or producing in the individuals so affected a defect in structure very different from that ordinarily found in their species; for example, ectromelia or cyclopia [43, *I, 33, 39–49*].

The interest of such a classification lies in the fact that it utilizes two different principles of discrimination and hierarchy: anomalies are arranged in terms of their increasing complexity and increasing seriousness. The simplicity–complexity relationship is purely objective. It goes without saying that a cervical rib is a simpler anomaly than ectromelia or hermaphroditism. The slight–serious relationship has a less clear-cut logical character. Undoubtedly the gravity of anomalies is an anatomical fact; the criterion of the anomaly's gravity lies in the *importance* of the organ as far as its physiological or anatomical connections are concerned [43, *I, 49*]. For the naturalist importance is an objective idea, but it is essentially a subjective one in the sense that it includes a reference to the life of a living being, considered fit to qualify this same life according to what helps or hinders it. This is so true that Geoffroy Saint-Hilaire added a third principle of classification (a physiological one) to the first two (complexity, gravity), that is, the relationship between anatomy and the exercise of functions (obstacle), and then a fourth, which is patently psychological, the introduction of the idea of a *harmful* or *disturbing* influence on the exercise of functions [43, *I, 38, 39, 41, 49*]. If one were tempted to accord this last principle only a subordinate role, let us reply that the case of heterotaxies emphasizes on the contrary both its precise meaning and considerable biological value. Geoffroy Saint-Hilaire created this term to designate modifications in the inner organization, that is, in the relations of the viscera without modification of the functions and external appearance. Until then these cases had not been studied much and constituted a gap in anatomical language. This should not be surprising, although it is difficult to imagine the possibility of a complex anomaly which not only does not obstruct the smallest function but also does not even produce the slightest deformity. "An individual affected by heterotaxy can enjoy very robust health; he can live a very long time; and often it is only after his death that the presence of anomaly is noticed, of which he himself had been unaware" [43, *I, 45, 46*]. This amounts to saying that the anomaly is ignored insofar as there is no manifestation of it in the order of vital values. Thus, even a scientist acknowledges that an anomaly is known to science only if it is first perceived in the consciousness, in the form of an obstacle to the performance of functions, or discomfort or harmfulness. But the sensation of obstacle, discomfort or harmfulness is a sensation which must be termed normative since it involves the even unconscious reference to a function and to an impulse to the completeness of their exercise. Finally, in order to be able to speak of an anomaly using scientific language, a being must have appeared to himself or to another as abnormal in the albeit unformulated language of the living. As long as the anomaly has no functional repercussions experienced consciously by the individual, in the case of man, or ascribed to life's dynamic polarity in every other living thing, the anomaly is either ignored (in the case of heterotaxies) or constitutes an indifferent *variety*, a variation on a specific theme; it is an irregularity like the negligible irregularities found in objects cast in the same mold. It might form the subject of a special chapter in natural history, but not in pathology.

On the other hand, if we assume that the history of anomalies and teratology are a necessary chapter in the biological sciences, expressing the originality of these sciences – for there is no special science of chemical or physical anomalies – it is because a new point of view can appear in biology and carve out new territory there. This point of view is that of vital *normativity*. Even for an amoeba, living means preference and exclusion. A digestive tract, sexual organs, constitute an organism's behavioral norms. Psychoanalytic language is indeed right to give the name *poles* to the natural orifices of ingestion and excretion. A function does not work indifferently in several directions. A need places the proposed objects of satisfaction in relation to propulsion and repulsion. There is a dynamic polarity of life. As long as the morphological or functional variations on the specific type do not hinder or subvert this polarity, the anomaly is a tolerated fact; in the opposite case the anomaly is felt as having negative vital value and is expressed as such on the outside. Because there are anomalies which are experienced or revealed as an organic disease, there exists first an affective and then a theoretical interest in them. It is because the anomaly has become pathological that it stimulates scientific study. The scientist, from his objective point of view, wants to see the anomaly as a mere statistical divergence, ignoring the fact that the biologist's scientific interest was stimulated by the normative divergence. In short, not all anomalies are pathological but only the existence of pathological anomalies has given rise to a special science of anomalies which, because it is science, normally tends to rid the definition of anomaly of every implication of a normative idea. Statistical divergences such as simple varieties are not what one thinks of when one speaks of anomalies;

instead one thinks of harmful deformities or those even incompatible with life, as one refers to the living form or behavior of the living being not as a statistical fact but as a normative type of life.

An anomaly is a fact of individual variation which prevents two beings from being able to take the place of each other completely. It illustrates the Leibnizean principle of indiscernibles in the biological order. But diversity is not disease; the *anomalous* is not the pathological. Pathological implies *pathos*, the direct and concrete feeling of suffering and impotence, the feeling of life gone wrong. But the pathological is indeed abnormal. Rabaud distinguishes between abnormal and sick because, following recent, incorrect usage, he makes "abnormal" the adjective of "anomaly" and in this sense speaks of abnormal sick people [97, *481*]; but as he distinguishes very clearly in other respects between disease and anomaly [97, *477*], following the criterion given for adaptation and viability, we see no reason to modify our distinctions of words and meanings.

Without doubt there is one way to consider the pathological normal, and that is by defining normal and abnormal in terms of relative statistical frequency. In a sense one could say that continual perfect health is abnormal. But that is because the word "health" has two meanings. Health, taken absolutely, is a normative concept defining an ideal type of organic structure and behavior; in this sense it is a pleonasm to speak of good health because health is organic well-being. Qualified health is a descriptive concept, defining an individual organism's particular disposition and reaction with regard to possible diseases. The two concepts, qualified descriptive and absolute normative, are so completely distinct that the same people will say of their neighbor that he has poor health or that he is not healthy, considering the presence of a fact the same as the absence of a value. When we say that continually perfect health is abnormal, we are expressing the fact that the experience of the living indeed includes disease. Abnormal means precisely nonexistent, inobservable. Hence it is only another way of saying that continual health is a norm and that a norm does not exist. In this misconstrued sense, it is obvious that the pathological is not abnormal. This is so little true that we can speak of the normal functions of organic defense and struggle against disease. As we have seen, Leriche asserts that pain is not in nature's plan, but we could say that disease is foreseen by the organism (Sendrail 106). With regard to the antibodies which are a defensive reaction against a pathological inoculation, Jules Bordet thinks that one can speak of normal antibodies which exist in normal serum acting electively on microbe and antigen, whose multiple specificities help assure the constancy of the organism's chemical characteristics by eliminating that which is not compatible with them [15, *6.16–14*]. But although disease may appear as foreseen, it is nonetheless true that it is like a state against which it is necessary to struggle in order to be able to go on living, that is, it is like an abnormal state in terms of the persistence of life which here serves as a norm. Hence in taking the word "normal" in its authentic sense we must set up an equation between the concepts of sick, pathological and abnormal.

Another reason for avoiding confusion between anomaly and disease is that human attention is not sensitized to each as being divergences of the same kind. An anomaly manifests itself in spatial multiplicity, disease, in chronological succession. It is a characteristic of disease that it interrupts a course; in fact it is critical. Even when the disease becomes chronic, after having been critical, there is a past for which the patient or those around him remain nostalgic. Hence we are sick in relation not only to others but also to ourselves. This is the case with pneumonia, arteritis, sciatica, aphasia, nephritis, etc. It is the characteristic of an anomaly that it is constitutional, congenital, even if its appearance is delayed with respect to birth and is contemporary only with the performance of a function – for example, in the congenital dislocation of the hip. The person with an anomaly cannot then be compared to himself. It could be pointed out here that the teratogenic interpretation of teratological characteristics, and better yet their teratogenetic explanation, allow the placement of the anomaly's appearance in embryological development and give it the significance of a disease. Once the etiology and pathology of an anomaly are known, the anomalous becomes pathological. Experimental teratogenesis provides some useful insights here [120]. But if this conversion of an anomaly into disease makes sense in the science of embryology, it makes no sense for the living being whose behavior in the environment, outside of the egg or uterus, is fixed at the outset by its structural characteristics.

When an anomaly is interpreted in terms of its effects in relation to the individual's activity and hence to the representation which develops from its value and destiny, an anomaly is an *infirmity*. Infirmity is a vulgar but instructive notion. One is born or one becomes infirm. It is the fact of becoming infirm which, interpreted as an irremediable breakdown, has repercussions for the fact of being born that way. For an invalid there exists in the end the possibility of some activity and an honorable social role. But a human being's forced limitation to a unique and invariable condition is judged pejoratively in terms of the normal human ideal, which is the potential and deliberate adaptation to every condition imaginable. It is the possible abuse of health which lies at the bottom of the value accorded to health just as it is the abuse of power which, according to Valéry, lies at the bottom of the love of power. Normal man is normative man, the being capable of establishing new, even organic norms. A single norm in life is felt privately, not positively. A man who cannot run feels injured, that is, he converts his injury into frustration, and although those around him avoid throwing up to him the image of his incapacity, just as sensitive children avoid running when a lame child is with them, the invalid feels sensitively by what restraint and avoidance on the part of his fellows each difference between him and them is apparently cancelled out.

What holds true for infirmity also holds true for certain states of *fragility* and *debility*, linked to a type of physiological divergence. This is the case with *hemophilia*, which is more an anomaly than a disease. All of the hemophiliac's functions are carried out like those of healthy individuals. But the hemorrhages are interminable, as if the blood were indifferent to its situation inside or outside the vessels. In short, the hemophiliac's life would be normal if animal life did not normally involve relations with an environment, relations whose risks in the form of injuries must be met by the animal in order to compensate for the disadvantages in feeding derived from the break with the inactive, vegetarian life; a break which, in other respects, particularly in terms of the development of consciousness, constitutes real progress. Hemophilia is a kind of anomaly with a possible pathological character because of the obstacle met here by an essential vital function, the strict separation of interior and exterior environment.

By way of summary: an anomaly can shade into disease but does not in itself constitute one. It is not easy to determine at what moment an anomaly turns into disease. Must the sacralization of the fifth lumbar vertebra be considered a pathological fact or not? There are certainly degrees of this malformation. Only the fifth vertebra must be termed sacralized when it is fused with the sacrum. Besides, in this case it rarely causes pain. Simple hypertrophy of a transverse apophysis, its more or less real contact with the sacral tubercle, are often deemed responsible for imaginary ills.

In short, we are dealing with anatomical anomalies of a congenital kind which become painful only later and sometimes never [101].

The problem of distinguishing between an anomaly – whether morphological like the cervical rib or sacralization of the fifth lumbar, or functional like hemophilia, hemeralopia or pentosuria – and the pathological state is not at all a clear one; but it is nevertheless quite important from the biological point of view because in the end it leads us to nothing less than the general problem of the variability of organisms and the significance and scope of this variability. To the extent that living beings diverge from the specific type, are they abnormal in that they endanger the specific form or are they inventors on the road to new forms? One looks at a living being having some new characteristic with a different eye depending on whether one is a fixist [*fixiste*] or a transformist. Understandably we haven't the slightest intention of dealing with such a problem here, though we cannot pretend to ignore it. When a drosophila with wings gives birth, through mutation, to a drosophila without wings or with vestigial wings, are we being confronted with a pathological fact or not? Biologists like Caullery, who do not admit that mutations are adequate for an understanding of the facts of adaptation and evolution, or like Bounoure, who dispute even the fact of evolution, insist on the subpathological or frankly pathological and even lethal character of most mutations. If they are not fixists like Bounoure [16] they at least agree with Caullery that mutations do not go beyond the framework of the species, since, despite considerable morphological differences, fertile crossbreeding is possible between control and mutant individuals [24, *414*]. It still seems indisputable that mutations can be the origin of new species. This fact was already well known to Darwin but it struck him less than individual variability. Guyénot thinks that it is the only presently known mode of hereditary variation, the only explanation, partial but unquestionable, of evolution [51]. Teissier and Philippe L'Héritier have demonstrated experimentally that certain mutations, which can seem disadvantageous in a species's usually appropriate environment, can become advantageous should certain conditions of existence vary. In a free and closed environment drosophila with vestigial wings are wiped out by drosophila with normal wings. But in an open environment the vestigial drosophila do not fly, feed constantly, and in three generations we see sixty percent vestigial drosophila in a mixed population [77]. This never happens in a closed environment. Let us not say normal environment because in the end, according to Geoffroy Saint-Hilaire, what is true of species is also true of environments: they are all that they must be as a function of natural laws, and their stability is not guaranteed. An open seashore environment is an indisputable fact, but this will be a more normal environment for wingless insects than for winged ones because those who do not fly are less likely to be eliminated. Darwin had noticed this fact, which was not taken seriously and which is confirmed and explained by the experiments reported above. An environment is normal because a living being lives out its life better there, maintains its own norm better there. An environment can be called normal with reference to the living species using it to its advantage. It is normal only in terms of a morphological and functional norm.

Teissier reports another fact which shows that, perhaps without looking for it, life, using the variation of living forms, obtains a kind of insurance against excessive specialization without reversibility, hence without flexibility, which is essentially a successful adaptation. In certain industrial districts in Germany and England the gradual disappearance of gray butterflies and the appearance of black ones of the same species has been observed. It was possible to establish that in these butterflies the black coloration was accompanied by an unusual vigor. In captivity the blacks eliminate the grays. Why isn't the same true in nature? Because their color stands out more against the bark of the trees and attracts the attention of birds. When the number of birds diminishes in industrial regions, butterflies can be black with impunity [111]. In short, this butterfly species, in the form of varieties, offers two combinations of opposing characteristics and they balance each other: more vigor is balanced by less security and vice versa. In each of the variations an obstacle has been circumvented, to use a Bergsonian expression, a powerlessness has been overcome. To the extent that circumstances allow one such morphological solution to operate in preference to another, the number of representatives of each variety varies, and a variety tends more and more toward a species.

Mutationism was first presented as a form of explanation for the facts of evolution, whose adoption by geneticists further reinforced the hostility shown toward every consideration of the influence of the environment. Today it seems that the appearance of new species must be placed at the intersection of innovations brought about by mutations and oscillations in the environment; and that a Darwinism rejuvenated by mutationism is the most flexible and comprehensive explanation of the fact of evolution – indisputable despite everything [56, *111*]. The species is the grouping of individuals, all of whom are different to some degree, whose unity expresses the momentary normalization of their relations with the environment, including other species, as Darwin had clearly seen. Taken separately, the living being and his environment are not normal: it is their relationship that makes them such. For any given form of life the environment is normal to the extent that it allows it fertility and a corresponding variety of forms such that, should changes in the environment occur, life will be able to find the solution to the problem of adaptation – which it has been brutally forced to resolve – in one of these forms. A living being is normal in any given environment insofar as it is the morphological and functional solution found by life as a response to the demands of the environment. Even if it is relatively rare, this living being is normal in terms of every other form from which it diverges, because in terms of those other forms it is *normative*, that is, it devalues them before eliminating them.

Hence, finally, we see how an anomaly, particularly a mutation, i.e., a directly hereditary anomaly, is not *pathological* because it is an anomaly, that is, a divergence from a specific type, which is defined as a group of the most frequent characteristics in their average dimension. Otherwise it would have to be said that a mutant individual, as the point of departure for a new species, is both pathological, because it is a divergence, and normal, because it maintains itself and reproduces. In biology the normal is not so much the old as the new form, if it finds conditions of existence in which it will appear normative, that is, displacing all withered, obsolete and perhaps soon to be extinct forms.

No fact termed normal, because expressed as such, can usurp the prestige of the norm of which it is the expression, starting from the moment when the conditions in which it has been referred to the norm are no longer given. There is no fact which is normal or pathological in itself. An anomaly or a mutation is not in itself pathological. These two express other possible norms of life. If these norms are inferior to specific earlier norms in terms of stability, fecundity, variability of life, they will be called pathological. If these norms in the same environment should turn out to be equivalent, or in another environment, superior, they will be called normal. Their normality will come to them from their normativity. The pathological is not the absence of a biological norm: it is another norm but one which is, comparatively speaking, pushed aside by life.

Here we have a new problem which leads us to the heart of our concerns and that is the relationship of the normal and the experimental. What physiologists after Bernard understand as normal phenomena are phenomena whose continuous exploration is possible thanks to laboratory equipment, and whose measured characteristics for any given individual in given conditions turn out to be identical to themselves; and, aside from some divergences of a clearly defined amplitude, identical from one individual to another in identical conditions. It would seem then that there is one possible definition of the normal, objective and absolute, starting from which every deviation beyond certain limits would logically be assessed as pathological. In what sense are laboratory standardization and mensuration appropriate to serve as the norm for the living being's functional activity considered outside the laboratory?

First of all, it should be pointed out that the physiologist, like the physicist and chemist, sets up experiments whose results he compares using this fundamental mental reservation that these data are valid "all other things being equal." In other words, other conditions would give rise to other norms. *The living being's functional norms* as examined in the laboratory are meaningful only within the framework of *the scientist's operative norms*. In this sense no physiologist would dispute the fact that he gives only a content to the concept of the biological norm but that in no case does he work out in what way such a concept is normative. Having admitted that some conditions are normal, the physiologist objectively studies the relations which actually define the corresponding phenomena, but he does not really objectively define which conditions are normal. Unless one admits that an experiment's conditions have no influence on the quality of the result – which is inconsistent with the care taken to determine them – one cannot deny the difficulty in assimilating experimental conditions with the normal ones of animal and human life, in the statistical as well as in the normative sense. If the abnormal or pathological is defined as a statistical divergence or as something unusual, as the physiologist usually defines it, it must be said, from a purely objective point of view, that the laboratory's conditions for examination place the living being in a pathological situation from which, paradoxically, one claims to draw conclusions having the weight of a norm. We know that this objection is very often directed at physiology, even in medical circles. Prus, in the same work from which we have already quoted a passage attacking Broussais's theories, states:

> Artificial diseases and the removal of organs practiced in experiments on living animals, lead to the same result [*as spontaneous diseases*]; however, it is important to point out that it would be wrong to proceed from services rendered by experimental physiology to favoring the influence physiology can exert on practical medicine.... When we irritate, puncture, cut the brain and cerebellum in order to learn the functions of these organs, or when we cut out a more or less considerable portion, the animal subjected to similar experiments is certainly as far removed as possible from the physiological state; it is seriously sick and what is called *experimental physiology* is obviously nothing other than a real *artificial pathology* which is similar to or creates diseases. Of course, physiology has its leading lights, and the names of Magendie, Orfila, Flourens will always have a place of honor in its annals; but these very figures offer an authentic and in some way material proof of everything this science owes to the science of disease [95, *L sqq*.].

It is to this kind of objection that Claude Bernard replied in the *Leçons sur la chaleur animale*:

> Certainly an experiment introduces disturbances into the or-

> ganism, but we must and can bear this in mind. We must restore the part of the anomalies which is due to them to the conditions in which we place the animal, and suppress the pain in animals as well as in man in order to remove causes for error brought about by suffering. But the very anesthetics we use have effects on the organism which can give rise to physiological modifications and new causes for error in our experiments' results [8, 57].

A noteworthy passage, which shows how close Bernard is to assuming that it is possible to discover a determinism of the phenomenon, independent of the determinism of the operation of knowledge; and how he is honestly obliged to acknowledge the alteration, in clearly unassignable proportions, to which knowledge subjects the known phenomenon because of the technical preparation it involves. When we glorify the contemporary theorists of wave mechanics for their discovery that observation interferes with the observed phenomenon, it happens that, as in other cases, the idea is a bit older than they are.

In the course of his research, the physiologist must come to grips with three kinds of difficulties. First he must be certain that the subject which is called normal in the experimental situation is identical with the subject of the same species in a normal, i.e., nonartificial situation. Then he must be certain of the similarity between the pathological state brought about by experiment and the spontaneous pathological state. Often the subject in the spontaneous pathological state belongs to a species other than the subject of the experimental pathological state. For example, without great precautions we cannot draw any conclusions about the diabetic human from Mering and Minkowski's dog, or Young's. Finally, the physiologist must compare the result of the two preceding comparisons. No one will question the breadth of the margin of uncertainty introduced by such comparisons. It is as vain to deny the existence of this margin as it is childish to question *a priori* the utility of such comparisons. In any case one understands the difficulty in realizing the canonical requirement of "all other things being equal." A convulsive crisis can be brought on by stimulating the cerebral cortex at the frontal ascendant, but it still is not epilepsy even if the electroencephalogram, after a succession of these crises, records superimposable curves. Four pancreases can be grafted simultaneously onto an animal without the animal experiencing the slightest hypoglycemic disorder comparable to that brought about by a small adenoma in the isles of Langerhans [53 *bis*]. Sleep can be induced by sleeping pills but according to A. Schwartz:

> It would be wrong to believe that sleep brought on by pharmacological means and normal sleep necessarily have an *exactly similar phenomenology* in these conditions. In reality the two cases are always different as the following examples prove: if, for example, the organism is under the influence of a *cortical* sedative, *paraldehyde*, the volume of urine *increases*, while in the course of normal sleep diuresis is usually reduced. The center of diuresis, initially liberated by the depressive action of the sedative on the cerebral cortex, is thus shielded from the subsequent inhibitory action of the sleep center.

It must be admitted that artificially inducing sleep by interfering with the nerve centers does not enlighten us as to the mechanism by which the hypnotic center is naturally put into operation by the normal factors of sleep [105, *23–28*].

If we may define the normal state of a living being in terms of a normal relationship of adjustment to environments, we must not forget that the laboratory itself constitutes *a new environment* in which life certainly establishes norms whose extrapolation does not work without risk when removed from the conditions to which these norms relate. For the animal or for man the laboratory environment is one possible environment among others. Certainly, the scientist is right in seeing in his apparatus only the theories which it materializes, to see in the products used only the reactions they allow; he is right in postulating the universal validity of these theories and these reactions, but for the living being apparatus and products are the objects among which he moves as in an unusual world. It is not possible that the ways of life in the laboratory fail to retain any specificity in their relationship to the place and moment of the experiment.

CHAPTER III

Norm and Average

It seems that in the concept of *average* the physiologist finds an objective and scientifically valid equivalent of the concept of normal or norm. Certainly the contemporary physiologist no longer shares Claude Bernard's aversion for every result of analysis or biological experiment expressed as an average, an aversion which perhaps originated in one of Bichat's texts:

> Urine, saliva, bile, etc., taken at random from this or that subject are analyzed and from their examination animal chemistry is born, whatever it may be. But this is not physiological chemistry; if I may say so, it is the cadaverous anatomy of fluids. Their physiology consists in the knowledge of innumerable variations undergone by the fluids as they follow the state of their respective organs [12, *art.* 7, §*1*].

Bernard is equally clear. According to him, the use of averages erases the essentially oscillatory and rhythmic character of the functional biological phenomenon. For example, if we look for the true number of heartbeats using the average of measurements taken several times on the same day from one given individual, "we will clearly have a false number." Hence this rule:

> In physiology average descriptions of experiments must never be given because the real relations of phenomena disappear in this average; when dealing with complex and variable experiments, we must study their different circumstances and then offer the most perfect experiment as type, which will always represent a true fact [6, *286*].

Research on average biological values has no meaning as far as the same individual is concerned: for example, the analysis of average urine over a 24-hour period is "the analysis of a urine which does not exist" since urine from the fasting state differs from that of digestion. This research is equally meaningless as far as individuals are concerned.

> The culmination [of this kind of experiment] was conceived by a physiologist who took urine from the urinal at the train station through which passed people of all nations, and believed he could thus produce the analysis of *average* European urine [6, *236*].

Without wishing to reproach Bernard for confusing research with its caricature and for loading a method with faults when responsibility for it lies with those who use it, we shall limit ourselves to maintaining that, according to him, the normal is defined as an ideal type in determined experimental conditions rather than as an arithmetical average or statistical frequency.

An analogous and again more recent attitude is that of Vendryès in his *Vie et probabilité* where Bernard's ideas on the constancy and regulations of the internal environment are systematically reexamined and developed. Defining physiological regulations as "the complex of functions which withstand chance" [115, *195*], or, if one wants functions which cause the living being's activity to lose its contingent and uncertain character (which would belong to it were the internal environment deprived of its autonomy vis-à-vis the external environment), Vendryès interprets the variations undergone by physiological constants – glycemia, for example – as divergences from an average, but an individual average. The terms divergence and average here have a probabilistic meaning. The greater the divergences the more improbable they are.

> I do not develop statistics on a certain number of individuals. I consider just one individual. The terms average value and divergence under these conditions are applied to the different values which the same component of the same individual's blood can assume in successive time periods [115, *33*].

But we do not think that Vendryès thereby eliminates the difficulty Bernard resolved by proposing the most perfect experiment as a type, that is, as a norm for comparison. In doing this, Bernard openly admitted that the physiologist brings to bear the norm of his own choosing in the physiology experiment and that he does not withdraw it. We do not think that Vendryès can proceed differently. He says that the *average* value of glycemia is 1% whereas [we know that] normally the rate of glycemia *is* 1%, but after eating or muscular work, glycemia undergoes positive or negative divergences from this average value. But assuming one effectively limits oneself to observing one individual, how does one conclude *a priori* that the individual chosen as the subject for the examination of variations of a constant represents the human type? Either one is a doctor – and this is apparently the case with Vendryès – and consequently qualified to diagnose diabetes; or one has learned nothing about physiology in the course of medical studies, and in order to learn the normal rate of one regulation one will look for the average of a certain number of results obtained from individuals placed in conditions as similar as possible. But in the end the problem is to know within what range of oscillations around a purely theoretical average value individuals will be considered normal.

A. Mayer [82] and H. Laugier [71] have dealt with this problem with great clarity and honesty. Mayer enumerates all the elements of contemporary physiological biometry: temperature, basal metabolism, blood gases, free heat, characteristics of the blood, rate of circulation, composition of the blood, reserves, tissues, etc. Now biological values allow a margin of variation. In order to represent a species we have chosen norms which are in fact constants determined by averages. The normal living being is the one who conforms to these norms. But must we consider every divergence abnormal?

> In reality the model is the product of statistics; most often it is the result of the calculations of averages. But the real individuals whom we meet diverge from these more or less and this is precisely in what their individuality consists. It would be very important to know what the divergences relate to and which divergences are compatible with extended survival. This should be known for the individuals of each species. Such a study is far from being done [82, *4.54–14*].

Laugier shows the difficulty of such a study dealing with man. He does it first by expounding Quetelet's theory of the *average man*, to which we shall return. The establishment of one of Quetelet's curves does not solve the problem of the normal for a given characteristic, for example, height. Guiding hypotheses and practical conventions are needed, allowing one to decide what value for heights, either toward the tall or the short, constitutes the transition from normal to abnormal. The same problem presents itself if we substitute a set of arithmetical averages with a statistical plan from which any individual diverges more or less, because statistics offer no means for deciding whether a divergence is normal or abnormal. Using a convention that reason itself seems to suggest, could one perhaps consider as normal the individual whose biometrical profile allows one to predict that, barring an accident, he will have a life span appropriate to his species? But the same questions reappear.

> In individuals who apparently die of senescence, we will find a very wide spread of life spans. Shall we take as the species's life span the average of these spans or the maximum spans reached by some rare individuals, or some other value? [71, *4.56–4*]

Moreover, this normality would not exclude other abnormalities: a certain congenital deformity can be compatible with a very long life. Strictly speaking, if the average state of the characteristic studied in the observed group can furnish a substitute for objectivity in the determination of a partial normality, the nature of the section about the average remains arbitrary; in any case all objectivity vanishes in the determination of a universal normality.

> Given the inadequacy of biometrical numerical data and the uncertainty as to where we are with regard to the validity of the principles to be used in establishing the dividing line between normal and abnormal, the scientific definition of normality, at the moment, seems beyond reach [*ibid.*].

Is it still more modest or, on the other hand, more ambitious to assert the logical independence of the concepts of norm and average and consequently the definitive impossibility of producing the full equivalent of the anatomical or physiological normal in the form of an objectively calculated average?

Starting with Quetelet's ideas and Halbwachs's very rigorous examinations of them, we intend to summarize the problem of the meaning and scope of biometric research in physiology. On the whole the physiologist who reviews its basic concepts is well aware that for him norm and average are two inseparable concepts. But average seems to him to be directly capable of objective definition and so he tries to join norm to it. We have just seen that this attempt at reduction runs into difficulties which are now, and undoubtedly always will be, insurmountable. Would it not be appropriate to turn the problem around and to ask whether the link between the two concepts couldn't be explained in terms of the subordination of average to norm? We know that biometry as applied to anatomy was first established by Galton's works, which generalized Quetelet's anthropometric procedures. In systematically studying the variations in human height, Quetelet had established and represented graphically the existence of a polygon of frequency showing an apex corresponding to the maximum ordinate, and a symmetry in terms of this ordinate for a characteristic measured in individuals of a homogeneous population. We know that the limit of a polygon is a curve and it was Quetelet himself who showed that the polygon of frequency tends toward a so-called "bell-shaped" curve which is the binomial or Gaussian error curve. By means of this relationship Quetelet expressly wanted to demonstrate that he recognized a given characteristic's individual variation (fluctuation) only in terms of that of an accident verifying the laws of chance, that is, the laws which express the influence of an unassignable multiplicity of nonsystematically oriented causes whose effects consequently tend to cancel out one another through progressive compensations. Now, to Quetelet this possible interpretation of biological fluctuations in terms of the calculation of probabilities seemed of the greatest metaphysical importance. According to him it meant there exists "a type or module" for the human race "whose different proportions could be easily determined" [96, *15*]. If this were not the case, if men differed from one another – with respect to height, for example – not because of the effect of accidental causes but because of the absence of a type with which they could be compared, no definite relationship could be established among all the individual measurements. On the other hand, if there is a type in terms of which divergences are purely accidental, a measured characteristic's numerical values, taken from many, many individuals, must be distributed according to a mathematical law and this is indeed what happens. In other respects, the greater the number of measurements carried out, the more the accidental disturbing causes will compensate and cancel out one another and the more clearly the general type will appear. But above all, from any large number of men whose height varies between determined limits, *those who come closest to the average height are the most numerous,* those who diverge from it the most are the least numerous. Quetelet called this human type – *from which the greater the divergence, the rarer it is* – the *average man*. When Quetelet is cited as the father of biometry, it is generally left unsaid that for him the average man is by no means an "impossible man" [96, *22*]. In a given region the proof of the average man's existence is found in the way the figures obtained for each dimension measured (height, head, arms, etc.) group themselves around the average by obeying the law of accidental causes. The average height in a given group is such that the largest of the subgroups formed of men of the same height is the set of men whose height comes closest to the average. This makes the typical average completely different from the arithmetical average.

When we measure the height of several houses we may get an average height but such that no house can be found whose own height approaches the average. In short, the existence of an average is, according to Quetelet, the indisputable sign of the existence of a regularity, interpreted in an expressly ontological sense:

> For me the principal idea is to cause the truth to prevail and to show how much man, without his knowledge, is subject to divine laws and with what regularity he realizes them. Moreover, this regularity is not peculiar to man: it is one of the great laws of nature belonging to animals as well as plants, and it will be surprising perhaps that it was not recognized sooner [96, *21*].

The interest of Quetelet's conception lies in the fact that in his notion of true average he identifies the ideas of *statistical frequency* and *norm*, for an average which determines that the greatest divergences are the most rare is really a norm. This is not the place to discuss the metaphysical foundation of Quetelet's thesis, but simply to argue that he distinguishes two kinds of averages: the arithmetical average or *median* and the true average; and that far from presenting the average as the empirical foundation of the norm with regard to human physical characteristics, he explicitly presents an ontological regularity which expresses itself in the average. If it should seem questionable to resort to God's will in order to understand the module for human height, this does not mean that no norm shows through in that average. And this seems to us to be what can be concluded from the critical examination to which Halbwachs subjected Quetelet's ideas [53].

According to Halbwachs, Quetelet is mistaken in considering the distribution of human heights around an average as a phenomenon to which the laws of chance can be applied. The first condition of this application is that phenomena, taken as combinations of elements of an unassignable number, are realizations which are completely independent of one another, so that no one of them exerts any influence on the one that follows. Now, constant organic effects cannot be assimilated with phenomena governed by the laws of chance. To do so is to admit that physical facts resulting from the environment and physiological facts related to the process of growth are arranged in such a way that each realization is independent of the others at an earlier, and at the same, time. This is untenable from the human point of view, where social norms interfere with biological laws so that the human individual is the product of a union subject to all kinds of customary and matrimonial legislative prescriptions. In short, heredity and tradition, habit and custom, are as much forms of dependence and interindividual connection as they are obstacles to an adequate utilization of the calculation of probabilities. Height, the characteristic studied by Quetelet, would be a purely biological fact only if it were studied in a set of individuals constituting a pure line, either animal or plant. In this case the fluctuations on both sides of the specific module would derive solely from the action of the environment. But in the human species height is a phenomenon inseparably biological and social. Even if height is a function of the environment, the product of human activity must be seen, in a sense, in the geographical environment. Man is a geographical agent and geography is thoroughly penetrated by history in the form of collective technologies. For example, statistical observation has made it possible to establish the influence of the draining of the Sologne marshes on the height of the inhabitants [89]. Sorre acknowledges that the average height of some human groups is probably raised under the influence of improved diet [109, *286*]. But we believe that if Quetelet made a mistake in attributing a value of a divine norm to the average of a human anatomical characteristic, this lies perhaps only in specifying the norm, not in interpreting the average as a sign of a norm. If it is true that the human body is in one sense a product of social activity, it is not absurd to assume that the constancy of certain traits, revealed by an average, depends on the conscious or unconscious fidelity to certain norms of life. Consequently, in the human species, statistical frequency expresses not only vital but also social normativity. A human trait would not be normal because frequent but frequent because normal, that is, normative in one given kind of life, taking these words *kind of life* in the sense given them by the geographers of the school of Vidal de la Blache [1845–1918; founder of French "human geography"].

This will appear even more obvious if, instead of considering an anatomical characteristic, we concentrate on a physiological one like longevity. Flourens, following Buffon, looked for a way to determine scientifically man's natural or normal life span, using and correcting Buffon's works. Flourens linked the life span to the specific duration of growth, whose term he defined in terms of the union of bones at their epiphyses.[24] "Man grows for twenty years and lives for five times twenty, that is, 100 years." That this normal human life span is neither the frequent nor the average duration is clearly specified by Flourens:

> Every day we see men who live 90 and 100 years. I am well aware that the number of those who reach that point is small when compared to the number of those who do not reach it, but in fact such ages are reached. And because they are sometimes reached, it is very possible to conclude that they would be reached more often, that they would be reached often if accidental and extrinsic circumstances, if disturbing causes did not get in the way. Most men die from disease; very few die, strictly speaking, of old age [39, *80–81*].

Metchnikoff also thinks that man can normally become a centenarian and that every old man who dies before 100 years of age is in theory a sick man.

The variations in man's average life span through the years (39 in 1865 and 52 in 1920 in France for males) are quite instructive. In order to assign a normal life to man, Buffon and Flourens considered him from the same biological perspective that they used for the rabbit or camel. But when we speak of an average life, in order to show it growing gradually, we link it to the action that man, taken collectively, exercises on himself. It is in this sense that Halbwachs deals with death as a social phenomenon, believing that the age at which death occurs results largely from working and hygienic conditions, attention paid to fatigue and diseases, in short, from social as much as physiological conditions. Everything happens as if a society had "the mortality that suits it," the number of the dead and their distribution into different age groups expressing the importance which the society does or does not give to the protraction of life [53, *94–97*]. In short, the techniques of collective hygiene which tend to prolong human life, or the habits of negligence which result in shortening it, depending on the value attached to life in a given society, are in the end a value judgment expressed in the abstract number which is the average human life span. The average life span is not the biologically normal, but in a sense the socially normative, life span. Once more the norm is not deduced from, but rather expressed in the average. This would be clearer still if, instead of considering the average life span in a national society taken as a whole, we broke this society down into classes, occupations, etc. We would see, of course, that the life span depends on what Halbwachs calls elsewhere the levels of life.

Undoubtedly it will be objected that such a conception is valid for superficial human characteristics for which there does exist, for the most part, a margin of tolerance where social diversities are in evidence, but that it certainly is not suitable for either fundamental human characteristics, which are essentially rigid such as glycemia, calcemia, or blood pH, or, generally speaking, for strictly specific characteristics in animals to which no collective technique offers any relative plasticity. Of course, we don't intend to maintain that anatomic–physiological averages express social norms and values in animals, but we do ask whether they wouldn't express vital norms and values. In the previous section we saw the example mentioned by G. Teissier of that butterfly species which oscillates between two varieties, tending to blend in with one or the other, depending on which of the two combinations that are compensated with contrasting characteristics the environment tolerates. We may well ask whether there wouldn't be a kind of general rule for the invention of living forms. Consequently, a very different meaning could be given to the existence of an average of the most frequent characteristics than that attributed to it by Quetelet. It would not express a specific stable equilibrium but rather the unstable equilibrium of nearly equal norms and forms of life temporarily brought together. Instead of considering a specific type as being really stable because it presents characteristics devoid of any incompatibility, it could be considered as being apparently stable because it has temporarily succeeded in reconciling opposing demands by means of a set of compensations. A normal specific form would be the product of a normalization between functions and organs whose synthetic harmony is obtained in defined conditions and is not given. This is almost what Halbwachs suggested in 1912 in his criticism of Quetelet:

> Why should we conceive of the species as a type from which individuals diverge only by chance? Why wouldn't its unity be the result of a duality of conformation, a conflict of two or a very small number of general organic tendencies which, all

> things considered, would balance each other out? What could be more natural than the expression of this divergence of its members' activities in terms of a regular series of divergences from the average in two different directions.... If these divergences were more numerous in one direction, this would indicate that the species tends to evolve in that direction under the influence of one or more constant causes [53, *61*].

As far as man and his permanent physiological characteristics are concerned, only a comparative human physiology and pathology – in the sense that there exists a comparative literature – of the various ethnic, ethical or religious, and technical groups and subgroups, which would take into account life's intricacy and its kinds and social levels, could furnish a precise answer to our hypotheses. It seems that this comparative human physiology, done from a systematic point of view, still remains to be written by a physiologist. Of course, there are compact compilations of biometrical data of anatomy and physiology concerning animal species as well as the human species separated into ethnic groups, for example the *Tabulae biologicae* [Junk, The Hague], but these are lists without any attempt at an interpretation of the results of the comparisons. By comparative human physiology we mean that kind of research best represented by the works of Eijkmann, Benedict and Ozorio de Almeida on basal metabolism and its relations with climate and race [Bibliography in 61, *299*]. But it happens that this gap has just been filled in part by the recent works of the French geographer, Sorre, whose *Les fondements biologiques de la géographie humaine* [The Biological Foundations of Human Geography] was drawn to our attention when the drafting of the essay was completed. We shall say something about this later, following a development which we want to leave in its primitive state, not so much out of concern for originality than as evidence of a convergence. Methodologically, the convergence by far prevails over the originality.

First of all, it will be agreed that in determining physiological constants by constructing averages obtained experimentally only within the laboratory framework, one would run the risk of presenting normal man as a mediocre man, far below the physiological possibilities of which men, acting directly and concretely on themselves or the environment, or obviously capable, even to the least scientifically informed observers. One may answer by pointing out that the frontiers of the laboratory have very much expanded since Claude Bernard; that physiology extends its jurisdiction over vocational guidance and selection centers and physical education institutes; in short, that the physiologist looks to the concrete man, not the laboratory subject in a very artificial situation; and that he himself determines the tolerated margins of variations with biometrical values. When A. Mayer writes:

> The very aim of the establishment of sports records is to measure the maximum activity of man's musculature [82, *4.54–14*],

we think of Thibaudet's witty remark:

> It is the record figures, not physiology, that answers the question: how many meters can a man jump? [*Le bergsonisme* (Paris, Editions de la nouvelle revue française, 1923) 1, 203].

In short, physiology would be only one sure and precise method for recording and standardizing the functional freedoms acquired or rather progressively mastered by man. If we can speak of normal man as determined by the physiologist, it is because normative men exist for whom it is normal to break norms and establish new ones.

As an expression of human biological normativity, not only do individual variations on the so-called civilized white man's common physiological "themes" seem interesting, but even more so are the variations of the themes themselves from group to group, depending on the types and levels of life, as related to life's ethical or religious attitudes, in short, the collective norms of life. In connection with this, Charles Laubry and Thérèse Brosse, thanks to the most modern recording techniques, have studied the physiological effects of the religious discipline which allows Hindu yogis almost complete mastery over the functions of vegetative existence. This mastery is such that it succeeds in regulating the peristaltic and antiperistaltic movements and in using the anal and vesical sphincters in every possible way, thus abolishing the physiological distinction between smooth and striated muscle systems. This mastery abolishes even the relative autonomy of the vegetative life. The simultaneous recording of pulse, respiration, electrocardiogram, and the measurement of basal metabolism have allowed one to establish that mental concentration, as it tends toward the fusion of the individual with the universal object, produces the following effects: accelerated heart rhythm, modification of the pulse's rhythm and pressure, and modification of the electrocardiogram: low generalized voltage, disappearance of waves, infinitesimal fibrillation on the isoelectric line, reduced basal metabolism [70, *1604*]. The key to the yogi's action on physiological functions, which seem least subject to the will, lies in breathing; it is breathing which is required to act on the other functions; by reducing it the body is placed "in the state of slowed existence comparable to that of hibernating animals" [*ibid.*]. To obtain a change in pulse rhythm from 50 to 150, an apnea [absence of respiration] of 15 minutes, an almost total suppression of cardiac contraction, certainly amounts to breaking physiological norms. Unless one chooses to consider such results pathological. But this is clearly impossible:

> If yogis are ignorant of the structure of their organs, they are indisputable masters of their functions. They enjoy a magnificent state of health and yet they have inflicted on themselves years of exercises which they couldn't have stood if they hadn't respected the laws of physiological activity [*ibid.*].

Laubry and Brosse conclude from such facts that we are in the presence of a human physiology which is very different from simple animal physiology: "The will seems to act as a pharmacodynamic test and for our superior faculties we glimpse an infinite power of regulation and order" [*ibid.*]. Whence these remarks of Brosse on the problem of the pathological:

> The problem of functional pathology, considered from the perspective of conscious activity related to the psychophysiological levels it uses, seems intimately connected with that of education. As the consequence of a sensory, active, emotional education, badly done or not done, it urgently calls for a reeducation. More and more the idea of health or normality ceases to appear as that of conformity to an outer idea (athlete in body, *bachelier* [*lycée* graduate] in mind). It takes its place in the relation between the conscious I and its psychophysiological organisms; it is relativist and individualist [17, *49*].

On these problems of physiology and comparative pathology we are forced to content ourselves with few documents, but, although their authors have followed dissimilar purposes, they lead one, surprisingly, to the same conclusions. Porak, who sought knowledge about the beginning of diseases in the study of functional rhythms and their disturbances, has demonstrated the relationship between kinds of existence and the curves of diuresis and temperature (slow rhythms), pulse and respiration (fast rhythms). Young Chinese between 18 and 25 have an average urinary discharge of 0.5 cm^3 per minute with oscillations from 0.2 to 0.7 while for Europeans this discharge is 1 cm^3 with oscillations from 0.8 to 1.5. Porak interprets this physiological fact in terms of the combined influences of geography and history in Chinese civilization. According to him, two out of this complex of influences are fundamental: the nature of the diet (tea, rice, young vegetables) and the nutritive rhythms determined by ancestral experience; – the mode of activity which more so in China than in the West respects the periodic development of neuromuscular activity. Western sedentary habits have a harmful effect on the rhythm of liquids. This disturbance does not exist in China, where the taste for walking "in the passionate desire to lose oneself in nature" has been preserved [94, *4–6*].

The study of respiratory rhythm (rapid rhythm) shows up variations in the need for activity related to development and to ankylosis. This need is itself related to natural or social phenomena which punctuate human work. Since the invention of agriculture, the solar day has framed the activity of most men. Urban civilization and the demands of a modern economy have disturbed the great physiological cycles of activity of which only traces remain. Onto these fundamental cycles are grafted secondary cycles. While changes in position determine secondary cycles in the variations of the pulse, it is the psychic influences which predominate in breathing. Breathing speeds up on awakening, as soon as the eyes open to the light:

> To open the eyes means that the attitude of the state of wakefulness is already being assumed; it means that the functional

> rhythms are already being oriented toward the deployment of neuromotor activity, and the supple respiratory function is ready to meet the outside world: it reacts immediately to the opening of the eyelids [94, *62*].

Because of the hematosis it guarantees, the respiratory function is so important for the explosive or sustained deployment of muscular energy that a very subtle regulation must determine instantaneously considerable variations in the volume of inhaled air. Respiratory intensity thus depends on the quality of our attacks or our reactions in our conflict with the environment. Respiratory rhythm is a function of our awareness of our situation in the world.

One would expect that Porak's observations would lead him to offer information about therapeutics and hygiene. And this is in fact what happens. Since physiological norms define less human nature than human habits as they relate to the kinds, levels and rhythms of life, every dietary rule must take these habits into account. Here is a good example of therapeutic relativism:

> Chinese women nurse their children during their first two years of life. After being weaned, the children will never drink milk again. Cow's milk is considered an unsuitable liquid, good only for pigs. I have often tried cow's milk with patients suffering from nephritis. Urinary ankylosis was produced immediately. By putting the patient on a diet of tea and rice, a good urinary crisis reestablished the eurhythmia [94, *99*].

As for the causes of functional diseases, if considered at their onset, they are almost all disturbances of rhythms, arhythmias stemming from fatigue or overwork, that is, from any exercise exceeding the proper adjustments of the individual's needs to the environment [94, *86*].

> It is impossible to maintain a type within his margin of functional availability. I believe the best definition of man would be an insatiable being, i.e., one who always exceeds his needs [94, *89*].

Here is a good definition of health that prepares us to understand its relationship to disease.

When Marcel Labbé studied the etiology of nutritional diseases principally with regard to diabetes, he came to analogous conclusions.

> Nutritional diseases are not organic but functional diseases. . . . Defects in diet play an important role in the origin of nutritional disturbances. . . . Obesity is the most frequent and the simplest of these diseases created by the *soft upbringing* [*éducation morbide*] provided by parents. . . . Most nutritional diseases are inevitable. . . . I am speaking above all of bad habits of life and diet which individuals must avoid and which parents already afflicted with nutritional disturbances must avoid passing on to their children [65, *10.501*].

We can only conclude that to consider the education of functions as a therapeutic measure, as Laubry and Brosse, Porak and Marcel Labbé do, is to admit that functional constants are habitual norms. What habit has made, habit unmakes and remakes. If diseases can be defined as defects in terms other than metaphorical, then physiological constants must be definable, other than metaphorically, as virtues in the old sense of the word, which blends virtue, power and function.

It must be said that Sorre's research on the relationship between man's physiological and pathological characteristics and climates, diets and biological environment, has an aim completely different from the works we have just cited. But what is noteworthy is that all these points of view are justified and their insights confirmed in Sorre's work. Men's adaptation to altitude and its hereditary physiological action [109, *51*]; the problems of the effects of light [109, *54*]; thermic tolerance [109, *58*]; acclimatization [109, *94*]; diet at the expense of a living environment created by man [109, *120*]; geographical distribution and the plastic action of diets [109, *245, 275*]; and the area of the extension of complex pathogens (sleeping sickness, malaria, plague, etc.) [109, *291*]: all these questions are treated with a great deal of precision, breadth and constant common sense. Certainly what interests Sorre above all is man's ecology, the explanation of the problems of human settlement. But in the end, as all these problems lead to problems of adaptation, we see how a geographer's work is of great interest for a methodological essay on biological norms. Sorre sees very clearly the importance of the cosmopolitanism of the human species for a theory of the relative instability of physiological constants: the importance of false adaptive equilibrium states to explain diseases or mutations; the relation of anatomical and physiological constants to collective diets, which he very judiciously qualifies as norms [109, *249*]; the irreducibility, to purely utilitarian reasons, of techniques for creating a really human ambience; the importance, in terms of the orientation of activity, of the indirect action of the human psyche on characteristics long considered natural such as height, weight, collective diatheses. By way of conclusion, Sorre is interested in showing that man, taken collectively, is searching for his "functional optima," that is, for values of each of the elements in his surroundings for which a particular function is best carried out. Physiological constants are not constants in the absolute sense of the term. For each function and set of functions there is a margin where the group or species capacity for adaptation comes into play. The optimal conditions thus determine a zone for human settlement where the uniformity of human characteristics expresses not only the inertia of a determinism but also the stability of a result maintained by an unconscious but real collective effort [109, *415–16*]. It goes without saying that we are pleased to see a geographer bringing the solidity of his results of analysis to bear in supporting our suggested interpretation of biological constants. Constants are presented with an average frequency and value in a given group which gives them the value of normal and this normal is truly the expression of a normativity. The physiological constant is the expression of a physiological optimum in given conditions among which we must bear in mind those which the living being in general, and *homo faber* in particular, give themselves.

Because of these conclusions we would differ somewhat from Pales and Monglond in interpreting their very interesting data on the rate of glycemia in African blacks [92 *bis*]. Out of 84 Brazzaville natives, 66% showed hypoglycemia; of these, 39% went from 0.90 g to 0.75 g and 27% were below 0.75 g. According to these authors the black must be generally considered as hypoglycemic. In any case the black withstands hypoglycemias which would be considered grave if not mortal in a European, without apparent disturbance and especially without either convulsions or coma. The causes of this hypoglycemia would have to be sought in chronic undernourishment, chronic and polymorphous intestinal parasitism and malaria.

> These states are on the border between physiology and pathology. From the European point of view they are pathological; from the indigenous point of view they are so closely linked to the black's habitual state that were it not for the comparative terms of the white, it could almost be considered physiological [92 *bis,* 767].

We definitely think that if the European can serve as a norm, it is only to the extent that his kind of life will be able to pass as normative. To Lefrou as well as to Pales and Monglond the black's indolence appears related to his hypoglycemia [76 *bis, 278*; 92 *bis, 767*]. These last authors say that the black leads a life in accordance with his means. But could it not just as well be said that the black has physiological means in accordance with the life he leads?

The relativity of certain aspects of anatomic and physiological norms and consequently of certain pathological disturbances as they relate to ways of life and knowledge of the world, is apparent not only in the comparison of ethnic and cultural groups which can be observed now, but also in the comparison of these present-day groups and earlier groups which have disappeared. Of course, paleopathology has even fewer documents at its disposal than paleontology or paleography, nevertheless the prudent conclusions which can be drawn from it deserve to be shown.

Pales, who has made a good synthesis of works of this kind in France, borrowed a definition of the paleopathological document from Roy C. Moodie, namely, every deviation from the healthy state of the body which has left a visible trace on the fossilized skeleton [92, *16*; in Pales 92 see a list of Moodie's works, and for a popularization see H. de Varigny, *La mort et la biologie*, Paris, Alcan, 1926]. If the sharpened flints and art of Stone Age men tell the story of their struggles, their works and their thought, their bones call to mind the history of their pains [92, *307*]. Paleopathology allows one to conceive of the pathological fact in human history as a fact of symbiosis, in the case of infectious diseases (and this concerns not only man but the living in general), and as a fact of the cultural level or kind of life, in the case of nutritional diseases.

The diseases suffered by prehistoric men turned up in very different proportions from those which diseases now offer for consideration. Vallois points out that in French prehistory alone eleven cases of tuberculosis turn up out of several thousand skeletons studied [113, *672*]. If the absence of rickets, a disease caused by vitamin D deficiency, is normal in an age where raw or barely cooked foods were consumed [113, *672*], the appearance of tooth decay, unknown to the first men, signifies civilization in terms of the consumption of starches and cooking food which brings in its wake the destruction of vitamins necessary for the assimilation of calcium [113, *677*]. Likewise, osteoarthritis was much more frequent in the Stone Age and subsequent epochs than it is now, and this must probably be attributed to an inadequate diet and a cold, humid climate, since its diminution in our day means better diet and a more hygienic way of life [113, *672*].

We can easily imagine the difficulty of a study which lacks all the diseases whose plastic or deforming effects failed to leave traces on the skeletons of fossil men or those dug up in the course of archaeological excavations. We can imagine the prudence necessary in drawing conclusions from this study. But to the extent that we can speak of a prehistoric pathology, we should also be able to speak of a prehistoric physiology just as we speak, without too much inaccuracy, of a prehistoric anatomy. Here again the relationship between the biological norms of life and the human environment seems to be both cause and effect of men's structure and behavior. Pales points out, with common sense, that if Boule could determine that the Man of La Chapelle-aux-Saints [a complete fossil skeleton found in 1908 in a cave in Corrèze, France] typifies the classical anatomy of the Neanderthal race, we [by contrast] could see in him without too much complacency, the most perfect type of pathological fossil man, suffering from alveolar pyorrhea, bilateral coxal–femoral osteoarthritis, cervical and lumbar spondylosis, etc. Yes, if we were to ignore the differences of cosmic milieu, technical equipment and way of life which make the abnormal of today the normal of yesterday.

If it seems difficult to dispute the quality of the observations used above, one might want to question the conclusions to which they lead concerning the physiological significance of functional constants interpreted as habitual norms of life. By way of response, it should be pointed out that these norms are not the product of individual habits which a certain individual could take or leave as he pleased. If we admit man's functional plasticity, linked in him to vital normativity, we are not dealing with either a total and instantaneous malleability or a purely individual one. To propose, with all suitable reservations, that man has physiological characteristics related to his activity, does not mean allowing every individual to believe that he will be able to change his glycemia or basal metabolism by the Coué method [of autosuggestion] or even by emigrating. What the species has worked out over the course of millennia does not change in a matter of days. Voelker has shown that basal metabolism does not change as one goes from Hamburg to Iceland. Benedict makes the same point concerning the moving of North Americans to subtropical regions. But Benedict has ascertained that the metabolism of Chinese permanently residing in the United States is lower than the American norm. Generally speaking, Benedict has established that [aboriginal] Australians (Kokatas) have a lower metabolism than whites of the same age, weight and height living in the United States, that inversely Indians (Mayas) have a higher metabolism with a slowed pulse and permanently lowered arterial pressure. We can conclude with Kayser and Dontcheff:

> It seems a proven fact that with man the climatic factor has no *direct* effect on metabolism; it is only in a very progressive manner that climate, by modifying the mode of life and allowing the consolidation of special races, has any lasting action on basal metabolism [62, *286*].

In short, to consider the average values of human physiological constants as the expression of vital collective norms would only amount to saying that the human race, in inventing kinds of life, invents physiological behaviors at the same time. But are the kinds of life not imposed? The works of the French school of human geography [Sorre and Vidal de la Blache] have shown that there is no geographical destiny. Environments offer man only potentialities for technical utilization and collective activity. Choice decides. Let us be clear that it is not a question of an explicit and conscious choice. But from the moment several collective norms of life are possible in a given milieu, the one adopted, whose antiquity makes it seem natural, is, in the final analysis, the one chosen.

In certain cases, however, it is possible to show the influence of an explicit choice on the direction of some physiological behavior. This is the lesson which emerges from observations and experiments related to temperature oscillations in the homeothermic animal, to circadian rhythm.

The works of Kayser and his collaborators on the pigeon's circadian rhythm established that the day and night variations in the central temperature of the homeothermic animal are a phenomenon of vegetative life subordinated to relational functions. The nocturnal reduction of exchanges is the effect of the suppression of light and sound stimulants. The circadian rhythm disappears in a pigeon made blind experimentally and isolated from his normal brethren. The reversal of the order in the light–dark succession reverses the rhythm after a few days. The circadian rhythm is determined by a conditioned reflex maintained by the natural alteration of day and night. As for the mechanism, it does not consist in a nocturnal hypoexcitability of the thermoregulatory centers but in the supplementary production by day of an amount of heat which is added to the calorification, evenly regulated day and night by the thermoregulatory center. This heat depends on stimulation coming from the environment as well as on the temperature: it increases with cold. When all heat produced by muscular activity is set aside, the rise which gives circadian temperature its rhythmic aspect must be related only to the increase in posture tonus by day. The homeothermic animal's circadian temperature rhythm is the expression of a variation in attitude of the entire organism with respect to the environment. Even when at rest the animal's energy, if it is stimulated by the environment, is not entirely at its disposal, one part being mobilized in tonic attitudes of vigilance and readiness. The state of wakefulness is a behavior which, even without alarms, does not work without costs [60; 61; 62; 63].

These conclusions shed a great deal of light on some observations and experiments concerning man, results which have often seemed contradictory. Mosso, on the one hand and Benedict on the other, were unable to show that the normal temperature curve depends on environmental conditions. But in 1907 Toulouse and Piéron stated that the inversion of the conditions of life (nocturnal activity and diurnal rest) brought about the complete inversion of the circadian temperature rhythm in man. How do we explain this contradiction? Benedict had observed subjects who were unaccustomed to nocturnal life and who at rest hours during the day led the normal life of their environment. According to Kayser, as long as experimental conditions do not equal those of a complete inversion of the mode of life, it is not possible to demonstrate a dependence between the rhythm and the environment. The following facts confirm this interpretation. In the suckling, circadian rhythm appears progressively, parallel to the infant's psychic development. At the age of eight days the divergence in temperature is 0.09°, at five months, 0.37°, between two and five years, 0.95°. Certain authors, Osborne and Voelker, have studied circadian rhythm in the course of long trips and they state that this rhythm follows the local time exactly [61, *304–306*]. Lindhard points out that during a Danish expedition to Greenland in 1906–08, circadian rhythm followed local time and that as far north as 76°46′ an entire crew, as well as the temperature curve, succeeded in shifting to the twelve-hour "day." Complete reversal could not be obtained because of the persistence of normal activity.[25]

Here then is an example of a constant related to the conditions of activity, to a collective and even individual kind of life, whose relativity expresses norms of human behavior in terms of a reflex conditioned to variable disengagement. Human will and human technology can turn night into day not only in the environment where human activity unfolds, but also in the organism itself whose activity confronts the environment. We do not know to what extent other physiological constants, when analyzed, could appear in the same way as the effect of a supple adaptation of human behavior. What matters to us is less to furnish a provisional solution than to show that a problem deserves to be posed. In any case, in this example we think we are using the term "behavior" correctly. From the moment the conditioned reflex sets the cerebral cortex's activity into operation, the term "reflex" must not be taken in its strict sense. We are dealing with a global, not a segmented, functional phenomenon.

By way of summary, we think that the concepts of norm and average must be considered as two different concepts: it seems vain to try to reduce them to one by wiping out the originality of the first. It seems to us that physiology has better to do than to search for an objective definition of the normal, and that is to recognize the original normative character of life. The true role of physiology, of sufficient importance and difficulty, would then be to determine exactly the content of the norms to which life has succeeded in fixing itself without prejudicing the possibility or impossibility of eventually correcting these norms. Bichat said that animals inhabit the world while plants belong only to their place of origin. This idea is even more true of men than of animals. Man has succeeded in living in all climates; he is the only animal – with the possible exception of spiders – whose area of expansion equals the area of the earth. But above all he is the animal who, through technology, succeeds in varying even the ambience of his activity on the spot, thereby showing himself now as the only species capable of variation [114]. Is it absurd to assume that in the long run man's natural organs can express the influence of the artificial organs through which he has multiplied and still multiplies the power of the first? We are aware that the heredity of acquired characteristics seems to most biologists to be a problem which has been resolved in the negative. We take the liberty of asking ourselves whether the theory of the environment's action on the living being were not on the verge of recovering from long discredit.[26*] True, it could be objected that in this case biological constants would express the effect of external conditions of existence on the living being and that our suppositions concerning the normative value of [natural] constants would be deprived of meaning. They would certainly be so if variable biological characteristics expressed change of environment, as variations in acceleration due to weight are related to latitude. But we repeat that biological functions are unintelligible as observation reveals them to us, if they express only states of a material which is passive before changes in the environment. In fact the environment of the living being is also the work of the living being who chooses to shield himself from or submit himself to certain influences. We can say of the universe of every living thing what Reininger says of the universe of man: "Unser Weltbild ist immer zugleich ein Wertbild,"[27] our image of the world is always a display of values as well.

Chapter IV

Disease, Cure, Health

In distinguishing anomaly from the pathological state, biological variety from negative vital value, we have, on the whole, delegated the responsibility for perceiving the onset of disease to the living being himself, considered in his dynamic polarity. That is to say, in dealing with biological norms, one must always refer to the individual because this individual, as Goldstein says, can find himself "equal to the tasks resulting from the environment suited to him" [46, *265*], but in organic conditions which, in any other individual, would be inadequate for these tasks. Just like Laugier, Goldstein asserts that a statistically obtained average does not allow us to decide whether the individual before us is normal or not. We cannot start from it in order to discharge our medical duty toward the individual. When it comes to a supra-individual norm, it is impossible to determine the "sick being" (*Kranksein*) as to content. But this is perfectly possible for an individual norm [46, *265, 272*].

In the same way, Sigerist insists on the individual relativity of the biological norm. If we are to believe tradition, Napoleon had a pulse of 40 even when he was in good health! If, with 40 contractions a minute, an organism is up to the demands imposed on him, then he is healthy and the number of 40 pulsations, though truly aberrant in terms of the average number of 70, is normal for this organism.[28] Sigerist concludes, "we should not be content with establishing the comparison with a norm resulting from the average, but rather, insofar as it is possible, with the conditions of the individual examined" [107, *108*].

If the normal does not have the rigidity of a fact of collective constraint but rather the flexibility of a norm which is transformed in its relation to individual conditions, it is clear that the boundary between the normal and the pathological becomes imprecise. But this in no way leads us to continuity between a normal and a pathological identical in essence save for quantitative variations, nor to a relativity of health and disease so confusing that one does not know where health ends and disease begins. The borderline between the normal and the pathological is imprecise for several individuals considered simultaneously but it is perfectly precise for one and the same individual considered successively. In order to be normative in given conditions, what is normal can become pathological in another situation if it continues identical to itself. It is the individual who is the judge of this transformation because it is he who suffers from it from the very moment he feels inferior to the tasks which the new situation imposes on him. The children's nanny, who perfectly discharges the duties of her post, is aware of her hypotension only through the neurovegetative disturbances she experiences when she is taken on vacation in the mountains. Of course, no one is obliged to live at high altitudes. But one is superior if one can do it, for this can become inevitable at any time. A norm of life is superior to another norm when it includes what the latter permits and what it forbids. But in different situations there are different norms, which, insofar as they are different, are all equal, and so they are all normal. In this regard Goldstein pays a great deal of attention to the sympathectomy experiments carried out on animals by Cannon and his collaborators. The animals, whose thermoregulation has lost all its usual flexibility and who are incapable of struggling for their food or against their enemies, are normal only in laboratory surroundings where they are sheltered from the brutal variations and sudden demands of adapting to the environment [46, *276–277*]. However, this normal is not called truly normal. For it is normal for the nondomesticated and nonexperimentally prepared living being to live in an environment where fluctuations and new events are possible.

As a consequence we must say that the pathological or abnormal state does not consist in the absence of every norm. Disease is still a norm of life but it is an inferior norm in the sense that it tolerates no deviation from the conditions in which it is valid, incapable as it is of changing itself into another norm. The sick living being is normalized in well-defined conditions of existence and has lost his normative capacity, the capacity to establish other norms in other conditions. It has long been noted that in tubercular osteoarthritis of the knee, articulation is frozen in a faulty position (the so-called Bonnet position). Nélaton was the first to give it its still classic explanation:

> It is rare that the limb maintains its usual straightness. Indeed, in order to relieve their suffering, the sick instinctively put themselves in an intermediary position between flexion and extension, which causes the muscles to exert less pressure on the articular surfaces [88, *II, 209*].

The hedonic and consequently normative significance of pathological behavior is perfectly perceived here. Articulation realizes its maximum capacity under the influence of muscular contraction and struggles spontaneously against pain. The position is said to be *defective* only in relation to an articulation practice which admits of all the possible positions save anterior flexion. But beneath this fault there is another norm, in other anatomic and physiological conditions, which lies hidden.

The clinical observation of men with head wounds, systematically carried out during the 1914–1918 war, allowed Goldstein to formulate some general principles about neurological nosology which can be appropriately summarized here.

If it is true that pathological phenomena are the regular modifications of normal phenomena, the former can shed some light on the latter only on the condition that the original meaning of this modification has been grasped. We must begin first by understanding the pathological phenomenon as revealing a modified individual structure. One must always bear in mind the transformation of the sick person's personality. Without this one runs the risk of ignoring the fact that the sick person, even though capable of reactions similar to those previously possible to him, can arrive at these reactions by very different paths. These reactions, which are apparently equivalent to previous normal reactions, are not the residue of previous normal behavior; they are not the result of an impoverishment or diminution; they are not the normal mode of life minus something which has been destroyed: they are reactions which never turn up in the normal subject in the same form and in the same conditions [45].

In order to define an organism's normal state, *preferential behavior* must be taken into account; in order to understand disease, *catastrophic reaction* must be taken into account. By preferential behavior it must be understood that among all the reactions of which an organism is capable under experimental conditions, only certain ones are used as those preferred. This mode of life, which is characterized by a set of preferred reactions, is that in which the living being responds best to the demands of his environment and lives in harmony with it; it is that which includes the most order and stability, the least hesitation, disorder and catastrophic reactions [46, *24*; 49, *131–134*]. Physiological constants (pulse, arterial pressure, temperature, etc.) express this ordered stability of behavior for an individual organism in well-defined environmental conditions.

> Pathological phenomena are the expression of the fact that the normal relationships between organism and environment have been changed through a change of the organism, and that thereby many things which had been adequate for the normal organism are no longer adequate for the modified organism.
>
> Disease is shock and danger for existence. Thus a definition of disease requires a *conception of the individual nature as a starting point*. Disease appears when an organism is changed in such a way that, though in its proper, "normal" milieu, it suffers catastrophic reaction. This manifests itself not only in specific disturbances of performance, corresponding to the locus of the defect, but in quite general disturbances because, as we have seen, disordered behavior in any field coincides always with more or less disordered behavior of the whole organism [46; English edition, p. 432].

What Goldstein pointed out in his patients is the establishment of new norms of life by a reduction in the level of their activity as related to a new but *narrowed* environment. The narrowing of the environment in patients with cerebral lesions corresponds to their impotence in responding to the demands of the normal, that is, previous environment. In an environment which is not rigidly protected, these patients would know only catastrophic reactions; insofar as the patient does not succumb to the disease, he is concerned with escaping from the anguish of the catastrophic reactions. Hence the mania for order and the meticulousness of these patients, their downright taste for monotony and their attachment to a situation they know they can dominate. The patient is sick because he can admit of only one norm. To use an expression which has already been very useful to us, the sick man is not abnormal because of the absence of a norm but because of his incapacity to be normative.

With this view of disease we see how far we are from the conception of Comte or Bernard. Disease is a positive, innovative experience in the living being and not just a fact of decrease or increase. The content of the pathological state cannot be deduced, save for a difference in format, from the content of health; disease is not a variation on the dimension of health; it is a new dimension of life. However new these views may seem to a French public,[29] they must not make one forget that in neurology they are the outcome of a long and fertile evolution of ideas begun by Hughlings Jackson.

Jackson represents disease of the nervous system of the relational life as dissolutions of hierarchical functions. Every disease corresponds to a level in this hierarchy. In every interpretation of pathological symptoms the negative as well as the positive aspect must be considered. Disease is both deprivation and change. A lesion in the higher nervous system frees the lower regulatory and control centers. Lesions are responsible for the loss of certain functions, but the disturbances of existing functions must be attributed to the appropriate activity of henceforth insubordinate centers. According to Jackson, no fact can have a negative cause. Neither a loss nor an absence is sufficient to produce disturbance in sensory neuromotor behavior [38]. Just as Vauvenargues says that people should not be judged on the basis of what they don't know, so Jackson proposes this methodological principle which Head called the golden rule: "Take note of what the patient really understands and avoid terms such as amnesia, alexia, word deafness, etc." [87, *759*]. It means nothing to say that a sick man has lost his speech if one does not specify in what typical situation this lack is perceptible. If a so-called aphasic subject is asked: Is your name John? He answers: No. But if he is ordered: Say no, he tries and fails. The same word can be said if it has the value of an interjection and cannot be said if it is a value judgment. Sometimes the sick person can't pronounce the word but gets to the point with a periphrasis. Suppose, says Mourgue, that the sick person, unable to name common objects, says, when presented with an inkwell: "This is what I would call a porcelain pot for holding ink," does he have amnesia or not? [87, *760*]

Jackson's important point is that language and, generally speaking, every function of relational life, is capable of several uses, in particular, an intentional use and an automatic use. In intentional actions, there is a preconception and action is carried out under control; it is dreamed of before being effectively executed. With language, two moments in the elaboration of an intentionally and abstractly significant proposition can be distinguished: a subjective moment, when notions automatically come to mind and an objective moment, when they are intentionally arranged according to a propositional plan. A. Ombredane points out that the divergence varies between these two moments *depending on the languages*:

> If there are languages where this divergence is very pronounced, as we can see in the final position of the verb in German, there are also languages where it is less. Moreover, if we remember that for Jackson the aphasic can scarcely go beyond the order of the subjective moment of expression, we, like Arnold Pick, can admit that the gravity of the aphasic disorder varies ac-

> cording to the structure of the language in which the sick person tries to express himself [91, *194*].

In short, Jackson's conceptions must serve as an introduction to Goldstein's. The sick person must always be judged in terms of the situation to which he is reacting and the instruments of action which the environment itself offers him – language, in cases of language disturbances. There is no pathological disturbance in itself: the abnormal can be evaluated only in terms of a relationship.

But no matter how correct the relationship established between Jackson and Goldstein by Ombredane [91], Ey and Rouart [38], and Cassirer [22], we should not ignore their profound difference and Goldstein's originality. Jackson's is an evolutionist point of view and he admits that the hierarchical centers of the relational functions correspond to different evolutionary stages. The relation of functional dignity is also one of chronological succession: higher and later functions are identified. The posteriority of the higher functions explains their fragility and precariousness. As disease is dissolution, it is also regression. The aphasic or apraxic rediscovers a child's or even an animal's language or gestures. Disease, although it represents a change in what remains and is not just the loss of what one had possessed, creates nothing, it throws the sick person, as Cassirer says, "a step backward on the road mankind had to clear slowly by means of constant effort" [22, *566*]. Now if it is true, according to Goldstein, that disease is a narrowed mode of life, lacking in creative generosity because lacking in boldness, it is nevertheless true that for the individual, disease is a new life, characterized by new physiological constants and new mechanisms for obtaining apparently unchanged results. Hence this warning, already cited:

> One must *refrain from believing that the various attitudes possible*

> *in a sick person represent just one kind of residue of normal behavior*, what has survived destruction. The attitudes which have survived in the sick person *never arise in that form in the normal person*, not even at the lower stages of its ontogenesis or phylogenesis, as is too frequently assumed. Disease has given them particular forms which cannot be well understood unless one considers the morbid state [45, *437*].

If it is possible, in effect, to compare the gesticulation of a sick adult with that of a child, the essential likening of one to the other would lead to the possibility of symmetrically defining the child's behavior as that of a sick adult. This would be an absurdity because it ignores that eagerness which pushes the child to raise itself constantly to new norms, which is profoundly at variance with the care to conserve which directs the sick person in his obsessive and often exhausting maintenance of the only norms of life within which he feels almost normal, that is, in a position to use and dominate his own environment.

On this very point Ey and Rouart have grasped the inadequacy of Jackson's conception:

> With regard to the psychic functions, dissolution produces both a capacitary regression and an involution toward a lower level of personality evolution. Capacitary regression does not exactly reproduce a past stage but it comes close to it (language, perceptual disturbances, etc.). The involution of the personality, insofar as it is precisely totalitarian, cannot be absolutely likened to a historical phase of ontogenic or phylogenic development, for it bears the mark of capacitary regression, and furthermore, as a reactional mode of the personality *at the actual moment*, it cannot go back to a past reactional mode, not even if it is cut off from its higher circumstances. This explains

> why, for all the analogies drawn between delirium and the child's mentality or primitive mentality, we cannot conclude that they are identical [38, *327*].

Again it was Jackson's ideas that guided Delmas-Marsalet in interpreting results obtained in neuropsychiatric therapy using electric shock. But not content to distinguish negative disturbances in terms of deficiency and positive disturbances in terms of the liberation of the remaining parts as Jackson did, Delmas-Marsalet, like Ey and Rouart, insists on what disease shows up as abnormal, to put it exactly, as new. In a brain subjected to toxic, traumatic, infectious effects, modifications consisting in new connections from area to area, in different dynamic orientations, can appear. A whole cell, which is quantitatively unchanged, is capable of a new arrangement, of different "isomeric" connections as isomers in chemistry are composed in an identical universal formula, but certain chains of which are placed differently in relation to a common nucleus. From the therapeutic point of view, it must be admitted that after dissolving neuropsychic functions, the coma obtained by means of electric shock makes possible a reconstruction which is not necessarily the inverted reappearance of stages in the previous dissolution. The cure can just as well be interpreted as a change from one arrangement to another, as seen as a restitution of the initial state [33]. If we point out these very recent conceptions here, it is to show the extent to which the idea that the pathological cannot be linearly deduced from the normal, tends to assert itself. Those who would be put off by Goldstein's language and manner will go along with Delmas-Marsalet's conclusions precisely because of what we personally shall consider as their weakness, that is, the vocabulary and images of psychological atomism (building, quarry stone, arrangement, architecture, etc.) used to formulate them. But in spite of the language, their clinical integrity establishes facts worth considering.

One may perhaps wish to object that in expounding Goldstein's ideas and their relation to Jackson's, we are moving in the area of psychic rather than somatic disturbances and that we are describing failures in psychomotor utilization rather than alterations in functions which are, strictly speaking, physiological, and which constitute the point of view we had said we had especially wanted to assume. We could answer that we have tackled not only the exposition but also the reading of Goldstein last and that all of the examples of pathological facts we have used to support our hypotheses and propositions – for which Goldstein's ideas are an encouragement and not an inspiration – are borrowed from physiological pathology. But we prefer to set out new, indisputably physiological pathological works whose authors owe nothing to Goldstein as far as the tendencies of their research are concerned.

In neurology it had long been noted through clinical observation and experimentation that severing nerves involves symptoms which cannot be adequately understood solely in terms of anatomical discontinuity. During the 1914–1918 war, a body of facts concerned with secondary sensorimotor disturbances, following injuries and surgical operations, again attracted attention. Explanations of that time introduced anatomical substitutes, pseudorestoration, and as often happens, for want of something better, pithiatism. Leriche's great merit is that from 1919 on he systematically studied the physiology of nerve stumps and systematized his clinical observations under the name of "neuroglioma syndrome." Nageotte called the swollen stump, which is often very large, the amputation neuroma, made of axis cylinders and neuroglia formed at the central end of a severed nerve. Leriche was the first to see that the neuroma is the starting point for a reflex phenomenon and he localized the origin of this so-called reflex in the neurites spread through the central stump. The neuroglioma syndrome includes a privative aspect, the appearance, in short, of an unprecedented disturbance. Assuming that the sympathetic fibers are the ordinary path of excitation originating at the level of the neuroglioma, Leriche thinks that these excitations

> determine unusual vasomotor reflexes at the wrong time, which are almost always vasoconstrictive, and these are the reflexes which, by producing hypermyotonia of smooth fiber, determine a truly new disease at the periphery, juxtaposed to the sensory motor deficiency related to severing the nerve. This new disease is characterized by cyanosis, chill, edema, trophic disturbances, pain, etc. [74, *153*].

Leriche's therapeutic conclusion is that neuroglioma formation must be prevented, particularly by means of a nerve graft. The graft does not perhaps reestablish anatomical continuity but it does in some way set the extremity of the central end and it channels the neurites by pushing them to the upper end. A technique developed by Foerster can also be used which consists in binding the neurolemma and mummifying the stump with an injection of absolute alcohol.

A.G. Weiss, working along the same lines as Leriche, thinks still more clearly than the latter that, with regard to disease of the neuroglioma, it is appropriate and sufficient to suppress the neuroglioma right away without losing time in "miming" the reestablishment of anatomical continuity by means of a graft or suture. With this procedure an integral restitution in the area of the injured nerve cannot be expected with any assurance. But it is a matter of choosing. For example, in the case of elbow seizure, one must choose between waiting for *possible* improvement of the paralysis if restoration of nervous continuity is effected following a graft, or *immediately* procuring for the patient the use of one hand which will always be partially paralyzed but which will be capable of very satisfying functional agility.

Klein's histological studies can perhaps explain all these phenomena [119]. Whatever the modalities of detail observed according to the cases (sclerosis, inflammation, hemorrhage, etc.), every histological examination of neuromata shows one constant fact, namely the persistent contact established between the axis cylinders' neuroplasm and the proliferation, sometimes considerable, of neurolemmata. This verification authorizes a close relationship between the neuromata and the receptor endings of the general sensibility, constituted by the ending of the neurites proper and by the elements differentiated but always deriving from the neurolemmata. This close relationship would confirm Leriche's conceptions that the neuroglioma is indeed a starting point for unusual excitations.

Be that as it may, A.G. Weiss and J. Warter are justified in asserting:

> To an uncommon degree the disease of the neuroglioma goes beyond the framework of the simple, sensitive motor interruption and very often, because of its seriousness, it constitutes the essence of the infirmity. This is so true that if one somehow succeeds in freeing the patient from disturbances linked to the existence of the neuroglioma, the sensory motor paralysis which persists assumes a truly secondary aspect, often compatible with almost normal use of the affected member [118].

The example of neuroglioma disease seems to us perfectly suited to illustrate the idea that disease is not merely the disappearance of a physiological order but the appearance of a new vital order, an idea which is as much Leriche's – as we saw in the first part of this study – as Goldstein's and which could correctly justify the Bergsonian theory of disorder. There is no disorder, there is the substitution for an expected or loved order of another order which either makes no difference or from which one suffers.

But Weiss and Warter, in pointing out that a functional restitution, satisfying in the eyes of the patient and also his doctor, can be obtained without a *restitutio ad integrum* in the theoretically corresponding anatomical order, confirm Goldstein's ideas on cure in a way which is certainly unexpected for them. Goldstein says

> Thus, *being well* means to be capable of ordered behavior which may prevail in spite of the impossibility of certain performances which were formerly possible. But the new state of health is not the same as the old one.... Just as a definite condition as to contents belongs to the *former* state of normality, so also a definite condition as to contents belongs to the *new* normality; but of course the contents of both conditions differ. This conclusion, which follows as a matter of course from our concept of the organism which is also determined as to contents, becomes of the greatest importance for the physician's attitude towards those who have regained their health.... To become well again, in spite of defects, always involves a certain loss in the essential nature of the organism. This coincides with the reappearance of order. A *new individual norm* corresponds to this rehabilitation.
>
> How very important the regaining of order is for recuperation can be seen from the fact that the organism seems primarily to have the tendency to preserve, or gain, such capacities which make this possible. The organism first of all appears set on gaining constants anew. We may find in recovery (with re-

sidual defect) changes in various fields as compared to the former nature of the organism; but the behavior shows that the character of the performances is again "constant." We find constants in the bodily as well as in the mental field. For instance, as compared to the former behavior, we find a change in a pulse rate, blood pressure, sugar content of the blood, in thresholds, mental performances, etc., but this modification is one of *newly* formed constants in the respective fields. These new constants guarantee the new order. We can understand the behavior of the recuperated organism only if we consider this fact. We must not attempt to interfere with these new constants, because we would thus create new disorders. We have learned that fever is not always to be combated, but that an increase in temperature may be understood as one of those constants which are necessary to bring about the recovery. We have learned to treat quite similarly certain forms of increased blood pressure or certain psychological changes. There are many such alterations of constants which today we still attempt to remove for their alleged harmfulness, whereas it would be better not to interfere with them [46; English edition, pp. 437–38].

One would gladly emphasize here – as opposed to one way of citing Goldstein which gives the appearance of initiation into a hermetic or paradoxical physiology – the objectivity and even banality of his leading ideas. It is not only the observations of clinicians (who are unfamiliar with his theses) but also experimental verifications which go along the lines of his own research. Didn't Kayser write in 1932:

The areflexia observed after a transverse spinal section stems from the interruption of the reflex arc itself. The disappearance of the state of shock accompanied by the reappearance of the reflexes is not, strictly speaking, a reestablishment but rather the constitution of a new "reduced" individual. A new entity is created, "the spinal animal" (von Weizsaecker) [63 *bis, 115*].

In asserting that new physiological norms are not the equivalent of norms existing before the disease, Goldstein, on the whole, only confirms the fundamental biological fact that life does not recognize reversibility. But if life does not admit of reestablishments, it does admit of repairs which are really physiological innovations. The more or less large reduction of these innovation possibilities is a measure of the seriousness of the disease. As far as health in the absolute sense is concerned, it is nothing other than the initial boundless capacity to institute new biological norms.

The frontispiece of Vol. VI of the *Encyclopédie française*, "The Human Being," published under Leriche's direction, shows health in the guise of an athlete throwing weights. This simple image seems to us to be as fully instructive as all the pages following, which are devoted to describing the normal man. We now want to gather together all our reflections scattered throughout earlier explanations and critical examinations in order to outline a definition of health.

If we acknowledge the fact that disease remains a kind of biological norm, this means that the pathological state cannot be called abnormal in an absolute sense, but abnormal in relation to a well-defined situation. Inversely, being healthy and being normal are not altogether equivalent since the pathological is one kind of normal. Being healthy means being not only normal in a given situation but also normative in this and other eventual situations. What characterizes health is the possibility of transcending the norm, which defines the momentary normal, the possibility of tolerating infractions of the habitual norm and instituting new norms in new situations. In an environment and system of given requirements, one remains normal with one kidney. But one can no longer allow oneself the luxury of losing a kidney, one must take care of it and oneself. Commonsense medical prescriptions are so familiar that we don't look for deep meaning in them. And yet how distressing and difficult it is to obey the doctor who says: Take care of yourself! "It is very easy to say take care of myself but I have my household to run," said the mother of a family in a hospital consultation, who, in saying it, had no intention of being ironic, no idea of semantics.[30] A household is the contingency of a sick husband or child, a torn pair of pants which must be mended in the evening when the child is in bed since he has only one pair of pants, the long trip to the bakery for bread if the usual one is closed for breaking the law, etc. How difficult it is to take care of oneself when one lived without knowing at what time one ate, whether the stairs were steep or not, the hour of the last tram since, if it were past, one would go home on foot, even a long way.

Health is a margin of tolerance for the inconstancies of the environment. But isn't it absurd to speak of the inconstancy of the environment? This is true enough of the human social environment where institutions are fundamentally precarious, conventions revocable, and fashions as fleeting as lightning. But isn't the cosmic environment, the animal environment in general a system of mechanical, physical and chemical constants, made of invariants? Certainly this environment, which science defines, is made of laws but these laws are theoretical abstractions. The living creature does not live among laws but among creatures and events which vary these laws. What holds up the bird is the branch and not the laws of elasticity. If we reduce the branch to the laws of elasticity, we must no longer speak of a bird, but of colloidal solutions. At such a level of analytical abstraction, it is no longer a question of environment for a living being, nor of health nor of disease. Similarly, what the fox eats is the hen's egg and not the chemistry of albuminoids or the laws of embryology. Because the qualified living being lives in a world of qualified objects, he lives in a world of possible accidents. Nothing happens by chance, everything happens in the form of events. Here is how the environment is inconstant. Its inconstancy is simply its becoming, its history.

For the living being life is not a monotonous deduction, a rectilinear movement, it ignores geometrical rigidity, it is discussion or explanation (what Goldstein calls *Auseinandersetzung)* with an environment where there are leaks, holes, escapes and unexpected resistances. Let us say it once more. We do not profess indeterminism, a position very well supported today. We maintain that the life of the living being, were it that of an amoeba, recognizes the categories of health and disease only on the level of experience, which is primarily a test in the affective sense of the word, and not on the level of science. Science explains experience but it does not for all that annul it.

Health is a set of securities and assurances (what the Germans call *Sicherungen)*, securities in the present, assurances for the future. As there is a psychological assurance which is not presumption, there is a biological assurance which is not excess, and which is health. Health is a regulatory flywheel of the possibilities of reaction. Life is usually just this side of its possibilities, but when necessary it shows itself above its anticipated capacity. This is clear in inflammation defense reactions. If the fight against infection were instantaneously victorious, there would be no inflammation. If organic defenses were immediately forced, there would no longer be inflammation. If inflammation exists it is because the anti-infectious defense is at once surprised and mobilized. To be in good health means being able to fall sick and recover, it is a biological luxury.

Inversely, disease is characterized by the fact that it is a reduction in the margin of tolerance for the environment's inconstancies. In speaking of reduction we do not mean to fall subject to the criticism we gave of the conceptions of Comte and Bernard. This reduction consists in being able to live only in another environment and not merely in some parts of the previous one. This is what Goldstein saw very clearly. At bottom, popular anxiety in the face of the complications of disease expresses nothing but this experience. We are more concerned about the disease any given disease may plunge us into than about disease itself, for it is more a matter of one disease precipitating another than a complication of disease. Each disease reduces the ability to face others, uses up the initial biological assurance without which there would not even be life. Measles is nothing, but it's bronchial pneumonia that we dread. Syphilis is so feared only after it strikes the nervous system. Diabetes is not so serious if it is just glycosuria. But coma? gangrene? what will happen if surgery is necessary? Hemophilia is really nothing as long as a traumatism does not occur. But who isn't in the shadow of a traumatism, barring a return to intrauterine existence? If even then!

Philosophers argue as to whether the living being's fundamental tendency is to conserve or expand. Medical experience would indeed seem to bring to bear an important argument in the debate. Goldstein notes that the morbid concern to avoid situations which might eventually generate catastrophic reactions expresses the conservation instinct. According to him, this instinct is not the general law of life but the law of a withdrawn life. The healthy organism tries less to maintain itself in its present state and environment than to realize its nature. This requires that the organism, in facing risks, accepts the eventuality of catastrophic reactions.

The healthy man does not flee before the problems posed by sometimes sudden disruptions of his habits, even physiologically speaking; he measures his health in terms of his capacity to overcome organic crises in order to establish a new order [49].

Man feels in good health – which is health itself – only when he feels more than normal – that is, adapted to the environment and its demands – but normative, capable of following new norms of life. It is obviously not with the express intention of giving men this feeling that nature built their organisms with such prodigality: too many kidneys, too many lungs, too much parathyroid, too much pancreas, even too much brain, if human life were limited to the vegetative life.[31]* Such a way of thinking expresses the most naive fatalism. But it has always been so: man feels supported by a superabundance of means which it is normal for him to abuse. As opposed to some doctors who are too quick to see crimes in diseases because those affected committed some excess or omission somewhere, we think that the power and temptation to fall sick are an essential characteristic of human physiology. To paraphrase a saying of Valéry, we have said that the possible abuse of health is part of health.

In order to evaluate the normal and the pathological, human life must not be limited to vegetative life. If need be, a man can live with many malformations or ailments but he can make nothing of his life, or, at least, he can always make something of it and it is in this sense that if it represents adaptation to imposed circumstances, every state of the organism, insofar as it is compatible with life, ends up being basically normal. But this normality is payed for by renouncing all eventual normativity. Man, even physical man, is not limited to his organism. Having extended his organs by means of tools, man sees in his body only the means to all possible means of action. Thus, in order to discern what is normal or pathological for the body itself, one must look beyond the body. With a disability like astigmatism or myopia, one would be normal in an agricultural or a pastoral society but abnormal for sailing or flying. From the moment mankind technically enlarged its means of locomotion, to feel abnormal is to realize that certain activities, which have become a need and an ideal, are inaccessible. Hence we cannot clearly understand how the same man with the same organs feels normal or abnormal at different times in environments suited to man unless we understand how organic vitality flourishes in man in the form of technical plasticity and the desire to dominate the environment.

If we now move back from these analyses to the concrete feeling of the state they are trying to define, we will understand that for man health is a feeling of assurance in life to which no limit is fixed. *Valere*, from which value derives, means to be in good health in Latin. Health is a way of tackling existence as one feels that one is not only possessor or bearer but also, if necessary, creator of value, establisher of vital norms. Hence this seduction still exerted on our minds today by the image of the athlete, a seduction of which contemporary infatuation for organized sport seems to us to be merely a sad caricature.[32]

Chapter V

Physiology and Pathology

As a consequence of the preceding analyses, it seems that a definition of physiology as the science of the laws or constants of normal life would not be strictly exact for two reasons: first, because the concept of normal is not a concept of existence, in itself susceptible of objective measurement; and second, because the pathological must be understood as one type of normal, as the abnormal is not what is not normal, but what constitutes another normal. This does not mean that physiology is not a science. It is genuinely so in terms of its search for constants and invariants, its metrical procedures, and its general analytical approach. But it is easy to specify *how* physiology is a science in terms of its method, less easy to specify *of what*, in terms of its object. Shall we call it the science of the conditions of health? In our opinion this would already be preferable to the science of the normal functions of life since we have believed we must distinguish between the normal state and health. But one difficulty persists. When we think of the object of a science we think of a stable object identical to itself. In this respect, matter and motion, governed by inertia, fulfill every requirement. But life? Isn't it evolution, variation of forms, invention of behaviors? Isn't its structure historical as well as histological? Physiology would then tend toward history, which is not, no matter what you do, the science of nature. It is true that we are nonetheless struck by life's stable quality. In short, in order to define physiology, everything depends on one's concept of health. Raphael Dubois, who is, to our knowledge, the only nineteenth-century author of a work on physiology in which a not merely etymological or purely tautological definition of it is proposed, derives its meaning from the Hippocratic theory of *natura medicatrix*:

> The role of *natura medicatrix* is identified with that of the normal functions of the organism which are all more or less directly conservative or defensive. Physiology is the study of nothing other than the functions of living beings, or in other words, the normal phenomena of the living proteon or bioproteon [35, *10*].

Now if we agree with Goldstein that there is only a really conservative tendency in disease, that the healthy organism is characterized by the tendency to face new situations and institute new norms, we cannot be satisfied with such a view.

Sigerist, who tries to define physiology by understanding the significance of the first discovery which gave rise to it – Harvey's discovery of the circulation of the blood (1628) – proceeds in his usual fashion, which is to place this discovery within the intellectual history of civilization. Why did a functional conception of life appear then, not sooner, not later? Sigerist does not separate the science of life, born in 1628, from the general, let us say, philosophical conception of life which was then expressed in the individual's various attitudes toward the world. From the end of the sixteenth and the beginning of the seventeenth century the plastic arts first established the baroque style and liberated movement everywhere. The baroque artist, as opposed to the classical artist, sees in nature only what is uncompleted, potential, not yet circumscribed.

> Baroque man is not interested in what is, but what is on the way to being. The baroque is infinitely more than a style in art, it is the expression of a form of thought which at this time governs all areas of the human spirit: literature, music, fashion, the State, the mode of living, the sciences [107, *41*].

In establishing anatomy at the beginning of the sixteenth century men favored the living form's static, delimited aspect. What Wölfflin says of the baroque artist, that he sees not the eye but the gaze, Sigerist says of the physician at the beginning of the seventeenth century:

> He does not see the muscle but its contraction and the effect it produces. This is how *anatomia animata*, physiology, is born. The object of this science is movement. It opens the doors to the unlimited. Each physiological problem leads to the sources of life and permits an escape to infinity [*ibid*].

Harvey, though an anatomist, saw not form but movement in the body. His research is not based on the configuration of the heart but on observing the pulse and respiration, two movements that cease only with life. The functional idea in medicine is connected with Michelangelo's art and Galileo's dynamic mechanics [107, *42*].[33]

It seems to us, following earlier considerations on health, that it goes without saying that this "spirit" of nascent physiology must be kept in the definition of physiology as the science of the conditions of health. We have spoken on several occasions of the modes of life, preferring this expression in certain cases to the term behavior in order to emphasize better the fact that life is dynamic polarity. It seems to us that in defining physiology as the *science of the stabilized modes of life*, we are meeting almost all the demands stemming from our previous positions. On the one hand we are assigning to research an object whose identity to itself is that of habit rather than nature, but whose relative constancy is perhaps more exactly adequate to take into account the nonetheless fluctuating phenomena with which the physiologist is concerned. On the other hand, we reserve the possibility for life to go beyond the codified biological constants or invariants conventionally held as norms at a specific moment of physiological knowledge. In effect, modes can be established only after having been put to the test by disrupting an earlier stability. Finally, it seems to us that starting from the definition proposed, we are able to delimit correctly the relations between physiology and pathology.

There are two kinds of original modes of life. There are those which are stabilized in new constants but whose stability will not keep them from being eventually transcended again. These are normal constants with propulsive value. They are truly normal by virtue of their normativity. And there are those which will be stabilized in the form of constants, which the living being's every anxious effort will tend to preserve from every eventual disturbance. These are still normal constants but with repulsive value expressing the death of normativity in them. In this they are pathological, although they are normal as long as the living being is alive. In short, the moment physiological stability is ruptured in a period of evolving crisis, physiology loses its rights but it does not for all that lose the thread. It does not know in advance whether the new biological order will be physiological or not, but later on it will have the means to find once more among the constants those which it claims for its own. This will be the case, for example, if the environment is made to vary experimentally in order to learn whether the constants which are maintained can accommodate themselves or not without catastrophe to a fluctuation in the conditions of existence. This is, for example, the leading thread which allows us to understand the difference between immunity and anaphylaxis. The presence of antibodies in the blood is common to both forms of reactivity. But while immunity makes the organism insensible to an intrusion of microbes or toxins in the inner environment, anaphylaxis is an acquired supersensitivity to the penetration of specific, particularly protein, substances into the inner environment [104]. After a first modification (by infection, injection or intoxication) of the inner environment, a second break-in is ignored by the immunized organism, while in the case of anaphylaxis, it provokes a shock reaction of extreme gravity, very often fatal, so sudden that it has qualified the experimental injection which provokes it with the term *unleashing* [*déchaînante*], hence a typically catastrophic reaction. The presence of antibodies in blood serum is thus always normal, the organism having reacted by modifying its constants to a first aggression of the environment and being regulated by it, but in one case the normality is physiological, in the other, pathological.

According to Sigerist, Virchow defined pathology as a "physiology with obstacles" [107, *137*]. This way of understanding disease by deriving it from normal functions, thwarted by a foreign addition which complicates them without altering them, comes close to the ideas of Claude Bernard and proceeds from very simple pathogenic principles. We know, for example, how a heart or kidney is made, how blood or urine passes through them; if we imagine the ulcerating growths of endocarditis on the mitral valve or a stone in the renal pelvis, we are in a position to understand the pathogeny of symptoms such as heart murmur or pain radiating from nephretic colic. But perhaps there is confusion in this conception, of a pedagogical and heuristic kind. Medical teaching rightly begins with the anatomy and physiology of the normal man, starting from which the reason for certain pathological states can sometimes be easily deduced by acknowledging certain mechanical analogies, for example, in the circulatory system: cardiac liver, dropsy, edemas; in the sensory motor system: hemianopsia or paraplegia. It seems that the order of acquiring these anatomic and physiological correspondences has been inverted. First of all, it is the sick man who one day ascertained that "something was wrong"; he noticed certain surprising or painful changes in his morphological structure or behavior. Rightly or wrongly he called them to the attention of his doctor. The latter, alerted by his patient, proceeded to a methodical exploration of the patent symptoms and even more the latent symptoms. If the patient died, an autopsy was performed, all kinds of means were employed to look for certain peculiarities in all the organs, which were compared with the organs of individual dead men who had never shown similar symptoms. The clinical observation and the autopsy report were compared. Here is how pathology, thanks to pathological anatomy but also thanks to hypotheses or knowledge concerning functional mechanisms, has become a physiology with obstacles.

Now here is a professional oversight – perhaps capable of being explained by the Freudian theory of lapses and failed acts – which must be pointed out. The physician has a tendency to forget that it is the patients who call him. The physiologist has a tendency to forget that a clinical and therapeutic medicine, which was not always so absurd as one might think, preceded physiology. Once this oversight is remedied, we are led to think that it is the experience of an obstacle, first lived by a concrete man in the form of disease, which has given rise to pathology in its two aspects, clinical semiology and the physiological interpretation of symptoms. If there were no pathological obstacles there would be no physiology because there would be no physiological problems to solve. Summarizing the hypotheses we proposed in the course of examining Leriche's ideas, we can say that in biology it is the *pathos* which conditions the *logos* because it gives it its name. It is the abnormal which arouses theoretical interest in the normal. Norms are recognized as such only when they are broken. Functions are revealed only when they fail. Life rises to the consciousness and science of itself only through maladaptation, failure and pain. A. Schwartz, following Ernest Naville, points out the glaring disproportion between the place occupied by sleep in men's lives and the place accorded it in works of physiology [105], just as Georges Dumas points out that the bibliography on pleasure is minute compared to the plethora of works devoted to pain. This is because the essence of sleep and enjoyment is to let life go without calling it to account.

In the *Traité de physiologie normale et pathologique* [Treatise on Normal and Pathological Physiology] [1], Abelous credits Brown-Séquard with having founded endocrinology by determining in 1856 that cutting out the adrenal glands brought about the death of an animal. It seems that this is a fact which is sufficient in itself. The question is not asked as to how it occurred to Brown-Séquard to carry out the removal of the adrenal glands. In the ignorance of the adrenal glands' functions, this is not a decision that one reaches by deduction. No, but it is the reflection of an accident. And in fact Sigerist shows that it is clinical practice which stimulated endocrinology. In 1855 Addison described the disease which since then has carried his name and which he attributed to an attack on the adrenal glands [107, 57]. Starting from this, Brown-Séquard's experimental research is understood. In the same *Traité de physiologie* [112, *1011*], Tournade judiciously points out the relation between Brown-Séquard and Addison and relates this anecdote of great epistemological significance: in 1716 the Bordeaux Academy of Sciences proposed as the subject of a competition: "What are the adrenal glands used for?" Montesquieu, who was responsible for the report, concluded that no paper submitted could satisfy the

Academy's curiosity and added: "One day perhaps chance will accomplish everything that all the effort in the world could not."

To take another example from the same kind of research, all physiologists trace the 1889 discovery of the role of the pancreatic hormone in glucide metabolism to von Mering and Minkowski. But it is often not known that if these two researchers made a dog diabetic, as famous in pathology as Saint Roch's in hagiography, it was quite unintentional. It was in studying external pancreatic secretion and its role in digestion that the dog had its pancreas removed. Naunyn, in whose department the experiment took place, says that it was summer and the lab boy was struck by the unusual number of flies around the animal cages. Naunyn, acting on the principle that where there is sugar, there are flies, recommended that the dog's urine be analyzed. Von Mering and Minkowski, then, by means of the pancreatectomy, had brought into being a phenomenon analogous to diabetes [2]. Thus artifice makes clarification possible, but without premeditation.

Likewise we should think a moment about these words of Déjerine:

> It is almost impossible to describe precisely the symptoms of paralysis of the glossopharyngeal nerve: in effect physiology has not yet established exactly the motor distribution of this nerve and on the other hand isolated paralysis of the glossopharyngeal nerve is never observed, so to speak, in clinical practice. In reality the glossopharyngeal nerve is always injured with the pneumogastric nerve or the spinal nerve, etc. [31, *587*].

It seems to us that the first if not the only reason why physiology has not yet established the exact motor distribution of the glossopharyngeal nerve is precisely because this nerve gives rise to no isolated pathological syndrome. When I. Geoffroy Saint-Hilaire attributed the gap corresponding to heterotaxies in the teratological science of his time to the absence of every morphological or functional symptom, he gave evidence of very rare perspicacity.

Virchow's conception of the relationship between physiology and pathology is inadequate not only because it ignores the normal order of logical subordination between physiology and pathology, but also because it implies that disease creates nothing of its own accord. We have dealt too explicitly on this last point to come back to it again. But the two errors seem to us to be connected. It is because disease is allowed no biological norm of its own that nothing is expected from it for the science of the norms of life. An obstacle would only slow down or stop or divert a force or current without altering them. Once the obstacle is removed, the pathological would again become physiological, the earlier physiological. Now this is what we cannot admit, following either Leriche or Goldstein. The new norm is not the old norm. And as this capacity to establish new constants with the value of norm has seemed to us to be characteristic of the living being's physiological aspect, we cannot admit that physiology can be constituted before and independently of pathology in order to establish it objectively.

Today it is not thought possible to publish a treatise on normal physiology without a chapter devoted to immunity, to allergy. Knowledge of the latter phenomenon reveals to us that about 97% of white men show a positive skin-test to tuberculin, without all of them, however, being tubercular. And yet this is the famous mistake of Koch, who is the source of this knowledge. Having ascertained that the tuberculin injection in an already tubercular subject gives rise to serious accidents, while it is harmless for a healthy subject, Koch believed that in tuberculinization he had found an infallible diagnostic tool. But having also wrongly attributed to it a curative value, he obtained results whose sad memory was effaced only by their subsequent conversion into a precise diagnostic instrument and preventive detection, namely the skin-test ascribed to von Pirquet. Almost every time that someone says in human physiology: "Today we know that . . . " one would find by looking hard – and without wishing to diminish the role of experimentation – that the problem was posed and its solution often outlined by clinical practice and therapeutics and very frequently at the expense, biologically speaking, of the patient. Thus, if Koch discovered in 1891 the phenomenon which bears his name and from which arose the theory of allergy and the skin-test technique, Marfan, as early as 1886, relying on the rarity of the coexistence of tubercular bone localizations, such as coxalgia or Pott's disease, and phthisis, had the intuition, clinically speaking, that certain tubercular manifestations could determine an immunity for others. In short, in the case of allergy, a general phenomenon of which anaphylaxis is one type, we discern the transition from an ignorant physiology to a knowing physiology by means of clinical practice and therapeutics. Today an objective pathology proceeds from physiology but yesterday physiology proceeded from a pathology which must be called subjective and thereby certainly imprudent, but certainly bold, and thereby progressive. All pathology is subjective with regard to tomorrow.

Is it only with regard to tomorrow that pathology is subjective? In this sense all science, which is objective in terms of its method and object, is subjective with regard to tomorrow since, short of assuming it to be completed, many of today's truths will become yesterday's mistakes. When Bernard and Virchow, each on his own, aimed at establishing an objective pathology, the one in the form of a pathology of functional regulations, the other in the form of cellular pathology, they tended to incorporate pathology into the sciences of nature, to found it on the bases of law and determinism.[34*] It is this claim that we want to subject to examination. If it has not seemed possible to maintain the definition of physiology as the science of the normal, it seems difficult to admit that there can be a science of disease, that there can be a purely scientific pathology.

These questions of medical methodology have not stirred up much interest in France, neither on the part of philosophers nor on the part of physicians. To my knowledge Pierre Delbet's old article in the collection *De la méthode dans les sciences* [32], has had no successors. On the other hand, these problems have been treated with great consistency and care abroad, particularly in Germany. We plan to borrow an exposition of the conceptions of Ricker and Magdebourg and the controversies they provoked, as given by Herxheimer in his *Krankheitslehre der Gegenwart* [Contemporary Pathology] (1927). We are deliberately giving this exposition the form of a summary, paraphrased and cut from quotations from pages 6 to 18 of Herxheimer's book [55].[35]

Ricker expounded his ideas successively in the *Entwurf einer Relationspathologie* [Outline of a Pathology of Relations] (1905); *Grundlinien einer Logik der Physiologie als reiner Naturwissenschaft* [Fundamentals of a Logic of Physiology as Pure Science] (1912); *Physiologie, Pathologie und Medizin* [Physiology, Pathology, and Medicine] (1923); *Pathologie als Naturwissenschaft – Relationspathologie* [Pathology as Science – Pathology of Relations] (1924). He delimits the areas of physiology, pathology, biology and medicine. The sciences of nature are based on methodical observation and reflection on these observations with a view to explaining, that is, articulating the causal relations between sensible, physical processes taking place in men's environment to which men themselves belong as physical beings. This excludes the psychism of the object of the sciences of nature. Anatomy describes morphological objects, its results have no explanatory value in themselves, but acquire it through their connection with the results of other methods, thus contributing to the explanation of phenomena which are the object of an independent science, physiology.

> While physiology explores the course of these processes, which is more frequent, more regular, and which is therefore called normal, pathology (which has been artificially separated from physiology) is concerned with their rarer forms, which are called abnormal; it must likewise be subjected to scientific methods. Physiology and pathology, joined together as one science, which could only be called physiology, examine the phenomena in physical man, with a view toward theoretical, scientific knowledge (*La pathologie comme science naturelle* [Pathology as Natural Science], p. 321) [55, 7].

Physiology-pathology must determine the causal relations between physical phenomena, but as no scientific concept of life exists – apart from a purely diagnostic concept – it has nothing to do with aims or ends and consequently with values in relation to life. All teleology (certainly not only transcendent but also immanent) which starts from the organism's finality or is related to it, to the preservation of life, etc., consequently every value judgment, does not belong to the natural sciences, still less to physiology-pathology [*ibid.*].

This does not exclude the legitimacy of value judgments or practical applications. But the former are relegated to biology as part of the philosophy of nature, hence part of philosophy; and the latter are relegated to medicine and hygiene considered as applied, practical, teleological sciences with the task of utilizing, according to their aims, what has been explained: "The teleological thought of medicine rests on the judgments of causality of physiology and pathology which form the scientific basis of medicine"

[55, *8*]. Pathology, as pure science of nature, must provide causal knowledge, but not produce value judgments.

Herxheimer responds to these propositions of general logic by saying first that it is not customary to class biology within philosophy as Ricker does, because if one relies on the expositions of representatives of the philosophy of values such as Windelband, Münsterburg and Rickert, biology cannot be granted the right to use really normative values; it must then be ranked among the natural sciences. Furthermore, certain concepts, like those of movement, nutrition, generation, to which Ricker himself grants a teleological meaning, are inseparable from pathology, both for psychological reasons peculiar to the subject concerned with it and for reasons residing in the objects themselves with which it is concerned [55, *8*].

Indeed, on the one hand scientific judgment, even when related to value-free objects, remains an axiological judgment because it is a psychological act. From the purely logical or scientific point of view it can be "advantageous," according to Ricker himself, to adopt certain conventions or certain postulates. And in this sense we can admit with Weigert or Peters a finality of the living being's organization or functions. From this point of view, notions such as those of activity, adaptation, regulation and self-preservation – notions which Ricker would eliminate from science – are advantageously maintained in physiology and thus in pathology as well [55, *9*]. In short, as Ricker clearly saw, scientific thought finds in everyday language, the nonscientific language of the masses, a defective instrument. But as Marchand says, we are not thereby obliged to "suspect a teleological ulterior motive in each simply descriptive term." Everyday language is particularly inadequate in the sense that its terms often have an absolute meaning while in our thought we give them only a relative meaning. To say, for example, that a tumor has an autonomous existence does not mean that it is really independent of the paths, materials and modes of nutrition of the other tissues but that compared to these, it is relatively independent. Even in physics and chemistry we use terms and expressions with an apparently teleological significance; however, no one thinks that they really correspond to psychical acts [55, *10*]. Ricker asks that biological processes or relations not be deduced from qualities or capacities. The latter must be analyzed in partial processes and their reciprocal reactions must be ascertained. But he himself admits that where this analysis is unsuccessful – in the case of nerve excitability, for example – the notion of a quality is inevitable and can serve as a stimulant for the search for the corresponding process. In his mechanics of development (*Entwicklungsmechanik*) Roux is obliged to admit certain qualities or properties of the egg, to use notions of preformation, regulation, etc., and yet Roux's research revolves around the causal explanation of the normal and abnormal processes of development [55, *11–12*].

On the other hand, if one takes the point of view of the very object of research, one must verify a withdrawal of the pretensions of physicochemical mechanics not only in biology but even in physics and chemistry. In any case pathologists, who answer in the affirmative to the question as to whether the teleological aspect of biological phenomena must be retained, are numerous, notably Aschoff, Lubarsch, Ziehen, Bier, Hering, R. Meyer, Beitzke, B. Fischer, Hueck, Roessle, Schwarz. With regard to serious brain lesions such as in tabes [progressive emaciation] or general paralysis, Ziehen for example, asks to what extent it is a matter of destructive processes and to what extent it is a matter of defensive or restorative processes conforming to a purpose, even if they lack it [55, *12–13*]. Schwartz's essay on "La recherche du sens comme catégorie de la pensée médicale" [The Investigation of Meaning as a Category of Medical Thought] must also be mentioned. He designates causality as a category – in the Kantian sense – of physics: "According to physics the conception of the world is determined by the application of causality as a category to a matter which is measurable, dispersed, without quality." The limits of such an application begin where such a breaking up into parts is not possible, where in biology objects appear which are characterized by increasingly distinct uniformity, individuality and totality. The competent category here is that of "meaning" [*sens*]. "Meaning is, so to speak, the organ through which we grasp structure, the fact of having form, in our thought; it is the reflection of structure in the consciousness of the observer." To the notion of meaning, Schwarz adds that of purpose, though it belongs to another order of value. But they have analogous functions in the two areas of knowledge and becoming from which they derive common qualities:

> Thus we grasp the meaning of our own organization in its tendency to preserve itself and only an environmental structure which contains meaning allows us to see purposes in it. Thus, through the consideration of purposes, the abstract category of meaning is filled with a real life. Consideration of purposes (for example, as a heuristic method) still remains always provisory, a substitute, so to speak, while waiting for the object's abstract meaning to become accessible to us.

By way of summary, in pathology a teleological way of looking at things is no longer rejected in principle by the majority of present-day scientists, yet terms with a teleological content have always been used without people being aware of it [55, *15–16*]. Of course, taking biological purposes into consideration must not exempt research from causal explanation. In this sense the Kantian concept of finality is always relevant. It is a fact, for example, that removing the adrenal glands brings about death. To assert that the adrenal capsule is necessary for life is a biological value judgment which does not relieve one from inquiring in detail into the causes through which a useful biological result is obtained. But in supposing that a complete explanation of the adrenal glands' functions is possible, teleological judgment, which recognizes the vital necessity of the adrenal capsule, would still retain its independent value, precisely in consideration of its practical application. Analysis and synthesis make a whole without one substituting for the other. *We must be aware of the difference between the two conceptions* [55, *17*]. It is true that the term "teleology" has remained too charged with implications of a transcendental kind to be gainfully employed; "final" is already better; but what would be better still would be "organismic," perhaps, used by Aschoff because it clearly expresses the fact of being related to the totality. This mode of expression is suited to the present tendency in pathology and elsewhere to put the total organism and its behavior again into the forefront [55, *17*].

Certainly Ricker does not absolutely proscribe such considerations but he does want to eliminate them totally from pathology as science of nature in order to relegate them to the philosophy of nature which he calls biology, and as far as their practical application is concerned, to medicine. Now this point of view poses precisely the question of whether such a distinction is useful in itself. This has been almost unanimously denied, and, it seems, with reason. Thus Marchand writes:

> For it is indeed true that pathology is not merely a natural science as far as the object of its research is concerned, but it also has the task of exploiting the result of its research for practical medicine.

Heuck, referring to Marchand, says that this would be impossible without the valorization and teleological interpretation of processes refused by Ricker. Let's think about a surgeon. What would he say if a pathologist, after performing a biopsy of a tumor, were to answer in sending him his findings, that whether a tumor is malignant or benign is a question for philosophy, not pathology? What would be gained in the division of labor advocated by Ricker? To a greater extent practical medicine would not get the solid scientific terrain on which it could be based. Hence we cannot go along with Honigmann, who, while approving Ricker's ideas for pathology but rejecting them for the practitioner, already draws the conclusion that physiology-pathology and anatomy must be shifted from the Faculty of Medicine to the Faculty of Science. The result would be to condemn medicine to pure speculation and *deprive physiology-pathology of stimulants of the greatest importance.* Lubarsch took the right view of things when he said:

> The dangers for general pathology and pathological anatomy lie primarily in the fact that they would become too unilateral and too solitary; closer relations between them and clinical practice as existed when pathology had not yet become a specialty, would certainly be of greater advantage to both parties [55, *18*].

There is no doubt that in defining the physiological state in terms of the frequency, and the pathological state in terms of the rarity of the mechanisms and structures they offer for consideration, Ricker can legitimately conceive that both must depend on the same heuristic and explanatory treatment. As we never believed it necessary to admit the validity of a statistical criterion, we cannot admit that pathology is completely aligned with physiology and becomes *science* while remaining science of the *pathological*. In fact, all those who accept the reduction of healthy and pathological biological phenomena to statistical facts are led more or less rapidly to acknowledge this postulate, implied in this reduction, that (according to a dictum of Mainzer quoted by Goldstein): "there is no difference between healthy life and sick life" [46, *267*].

We have already seen when we examined Claude Bernard's theory in what precise sense such a proposition can be defended. The laws of physics and chemistry do not vary according to health or disease. But to fail to admit that from a biological point of view, life differentiates between its states means condemning oneself to be even unable to distinguish food from excrement. Certainly a living being's excrement can be food for another living being but not for him. What distinguishes food from excrement is not a physicochemical reality but a biological value. Likewise, what distinguishes the physiological from the pathological is not a physicochemical objective reality but a biological value. As Goldstein says, when we are led to think that disease is not a biological category, this should already make us question the premises from which we started:

> How is it thinkable that disease and health should not be biological concepts! If we disregard, for a moment, the complicated conditions in man, this statement is certainly not valid for animals, where disease so frequently decides whether the individual organism is "to be or not to be." Just think what detrimental part disease plays in the life of the undomesticated animal, i.e., the animal which does not benefit by the protection through man! If the science of life is supposed to be incapable of comprehending the phenomena of disease, one must doubt seriously the appropriateness of, and the truth in, the intrinsic categories of a science so construed [46; English edition, p. 430].

Ricker of course acknowledges biological values, but in refusing to incorporate values into the object of a science, he makes the study of these values part of philosophy. He has been reproached – justly in our and Herxheimer's opinion – for this inclusion of biology in philosophy.

How then to resolve this difficulty: if we look at it from the strictly objective point of view there is no difference between physiology and pathology; if we look for a difference between them have we left scientific ground?

We would propose the following considerations as elements of a solution:

1. In the strict sense of the term, according to French usage, the science of an object exists only if this object allows measurement and causal explanation, in short, analysis. Every science tends toward metrical determination through establishing constants or invariants.

2. This scientific point of view is an abstract point of view, it expresses a choice and hence a neglect. To look for what men's lived experience is in reality is to overlook what value it is capable of accepting for and by them. Before science it is technologies, arts, mythologies and religions which spontaneously valorize human life. After the appearance of science these same functions still exist but their inevitable conflict with science must be regulated by philosophy, which is thus expressly philosophy of values.

3. The living being, having been led, in his humanity, to give himself methods and a need to determine scientifically what is real, necessarily sees the ambition to determine what is real extend to life itself. Life becomes – in fact, it has become so historically, not having always been so – an object of science. The science of life finds that it has life as subject, since it is the enterprise of living men, and as object.

4. In seeking to determine the constants and invariants which

really define the phenomena of life, physiology is genuinely doing the work of science. But in looking for what is the vital significance of these constants, in qualifying some as normal and others as pathological, the physiologist does more – not less – than the strict work of science. He no longer considers life merely as a reality identical to itself but as polarized movement. Without knowing it, the physiologist no longer considers life with an indifferent eye, with the eye of a physicist studying matter; he considers life in his capacity as a living being through whom life, in a certain sense, also passes.

5. The fact is that the physiologist's scientific activity, however separate and autonomous he may conceive of it in his laboratory, maintains a more or less close, but unquestionable relationship with medical activity. It is life's setbacks which draw and have drawn attention to it. Knowledge always has its source in reflection on a setback to life. This does not mean that science is a recipe for processes of action but that on the contrary the rise of science presupposes an obstacle to action. It is life itself, through its differentiation between its propulsive and repulsive behavior, which introduces the categories of health and disease into human consciousness. These categories are biologically technical and subjective, not biologically scientific and objective. Living beings prefer health to disease. The physician has sided explicitly with the living being, he is in the service of life and it is life's dynamic polarity which he expresses when he speaks of the normal and the pathological. The physiologist is often a physician, always a living man, and this is why the physiologist includes in his basic concepts the fact that if the living being's functions assume modes all equally explicable by the scientist, they are not for this reason the same for the living being himself.

To summarize, the distinction between physiology and pathology has and can only have a clinical significance. This is the reason why, contrary to all present medical custom, we suggest that it is medically incorrect to speak of diseased organs, diseased tissues, diseased cells.

Disease is behavior of negative value for a concrete individual living being in a relation of polarized activity with his environment. In this sense, it is not only for man – although the terms pathological or malady, through their relation to *pathos* or *mal*, indicate that these notions are applied to all living beings through sympathetic regression starting from lived human experience – but for every living thing that there is only completely organic disease. There are diseases of the dog and the bee.

To the extent that anatomical and physiological analysis breaks the organism down into organs and elementary functions, it tends to place disease on the level of partial anatomical and physiological conditions of the total structure or behavior as a whole. Depending on the degree of subtlety in the analysis, disease will be placed at the organ level – and it is Morgagni – at the tissue level – and it is Bichat – at the cellular level and it is Virchow. But in doing this we forget that historically, logically and histologically we reached the cell by moving backward, starting from the total organism; and thought, if not the gaze [*le regard*], was always turned toward it. The solution to a problem posed by the entire organism, first to the sick man, later to the clinician, has been sought in the tissue or cell. To look for disease at the level of cells is to confuse the plane of concrete life, where biological polarity distinguishes between health and disease, with the plane of abstract science, where the problem gets a solution. We do not mean that a cell cannot be sick if by cell we mean an entire living thing, as for example a protist [unicellular organism], but we do mean that the living being's disease does not lodge in parts of the organism. It is certainly legitimate to speak of a sick leucocyte just as one has the right to consider the leucocyte outside of every relation to the reticulo-endothelial system and the conjunctive system. But in this case the leucocyte is considered as an organ and better as an organism in a defense and reaction situation vis-à-vis an environment. In fact, the problem of individuality is posed here. The same biological given can be considered as part or as whole. We suggest that it is as a whole that it can be called sick or not.

Cells of the renal or pulmonary or splenic parenchyma can be called sick today or sick with a certain disease by a certain anatomist or pathologist, who has perhaps never set foot in a hospital or clinic, only because these cells were removed, or they resemble ones which were removed, yesterday or a hundred years ago – it doesn't matter – by a practicing physician, clinician and therapist, from the cadaver or amputated organ of a man whose behavior he had observed. This is so true that Morgagni, the founder of pathological anatomy, in his fine epistle to the surgeon Trew at the beginning of his basic work, enunciates the formal obligation of anatomic pathological exploration to refer constantly to the anatomy of the normal living being, obviously, but also and above all to clinical experience [85]. Virchow himself, coming to Velpeau's aid in a famous discussion in which French micrographers argued against him for the specific character of the cancerous element, proclaimed that if the microscope is capable of serving clinical practice, it is up to clinical practice to enlighten the microscope [116]. It is true that Virchow has elsewhere and with the greatest clarity formulated a theory of disease of the parts [*maladie parcellaire*] which our preceding analyses tend to refute. Did he not say in 1895:

> It is my idea that the essence of disease is a modified part of the organism or a modified cell or modified aggregate of cells

> (or tissue or organ). . . . In reality every sick part of the body is in a parasitic relation to the rest of the healthy body to which it belongs and lives at the expense of the organism [23, *569*].

It seems that today this atomistic pathology has been abandoned and that disease is seen much more as a reaction of everything organic against the attack of an element than as an attribute of the element itself. It is precisely Ricker who, in Germany, is the great opponent of Virchow's cellular pathology.[36]* What he calls the "pathology of relations" is precisely the idea that disease does not exist at the level of the supposedly autonomous cell but consists for the cell in the relations above all with the blood and nervous system, that is, with an interior environment and a coordinating organ which make the organism's functioning a whole [55, *19*]. It does not matter that the content of Ricker's pathological theories seems arguable to Herxheimer and others; what is interesting is the spirit of his attack. In short, when we speak of objective pathology, when we think that anatomical and histological observation, the physiological test, the bacteriological examination, are methods which enable the diagnosis of disease to have scientific significance – even, according to certain people, in the absence of all clinical inquiry and exploration – we are, in our opinion, victims of the most serious philosophical and, therapeutically speaking, sometimes the most dangerous confusion. A microscope, a thermometer, a culture medium know no medicine which the physician would not know. They give a result. This result has no diagnostic value in itself. In order to reach a diagnosis the sick person's behavior must be observed. It is then discovered that one who has a Löffler bacillus in his pharynx does not have diphtheria. On the other hand, for another man, a thorough and very accurately carried out clinical examination makes one think of Hodgkin's disease when the pathological examination of a biopsy reveals the existence of a thyroid tumor.

In pathology the first word historically speaking and the last word logically speaking comes back to clinical practice. Clinical practice is not and will never be a science even when it uses means whose effectiveness is increasingly guaranteed scientifically. Clinical practice is not separated from therapeutics, and therapeutics is a technique for establishing or restoring the normal whose end, that is, the subjective satisfaction that a norm is established, escapes the jurisdiction of objective knowledge. One does not scientifically dictate norms to life. But life is this polarized activity of debate with the environment, which feels normal or not depending on whether it feels that it is in a normative position or not. The physician has sided with life. Science serves him in fulfilling the duties arising from that choice. The doctor is called by the patient.[37] It is the echo of this pathetic call which qualifies as pathological all the sciences which medical technology uses to aid life. Thus it is that there is a pathological anatomy, a pathological physiology, a pathological histology, a pathological embryology. But their pathological quality is an import of technical and thereby subjective origin. There is no objective pathology. Structures or behaviors can be objectively described but they cannot be called "pathological" on the strength of some purely objective criterion. Objectively, only varieties or differences can be defined with positive or negative vital values.

Conclusion

In Part One we looked into the historical origins and analyzed the logical implications of the principle of pathology, so often still invoked, according to which the morbid state in the living being is only a simple quantitative variation of the physiological phenomena which define the normal state of the corresponding function. We think we have established the narrowness and inadequacy of such a principle. In the course of the discussion and in the light of the examples presented, we think we have furnished some critical arguments to support proposals of method and doctrine which form the object of Part Two and which we shall summarize as follows:

Types and functions can be qualified as normal with reference to the dynamic polarity of life. If biological norms exist it is because life, as not only subject to the environment but also as an institution of its own environment, thereby posits values not only in the environment but also in the organism itself. This is what we call biological normativity.

Without being absurd, the pathological state can be called normal to the extent that it expresses a relationship to life's normativity. But without being absurd this normal could not be termed identical to the normal physiological state because we are dealing with other norms. The abnormal is not such because of the absence of normality. There is no life whatsoever without norms of life, and the morbid state is always a certain mode of living.

The physiological state is the healthy state, much more than the normal state. It is the state which allows transition to new norms. Man is healthy insofar as he is normative relative to the fluctuations of his environment. According to us, physiological constants have, among all the possible vital constants, a propulsive value. The pathological state, on the other hand, expresses the reduction of the norms of life tolerated by the living being, the precariousness of the normal established by disease. Pathological constants have a repulsive and strictly conservative value.

Cure is the reconquest of a state of stability of physiological norms. It is all the closer to health or disease as this stability is more or less open to eventual change. In any case no cure is a return to biological innocence. To be cured is to be given new norms of life, sometimes superior to the old ones. There is an irreversibility of biological normativity.

The concept of norm is an original concept which, in physiology more than elsewhere, cannot be reduced to an objective concept determinable by scientific methods. Strictly speaking then, there is no biological science of the normal. There is a science of biological situations and conditions *called* normal. That science is physiology.

The attribution of a value of "normal" to constants whose physiology scientifically determines the content, expresses the relation of the science of life to life's normative activity and, as far as the science of human life is concerned, to biological techniques of production and establishment of the normal, and more especially to medicine.

It is with medicine as with all other technologies. It is an activity rooted in the living being's spontaneous effort to dominate the environment and organize it according to his values as a living being. It is in this spontaneous effort that medicine finds its meaning, if not at first all the critical clarity which renders it infallible. Here is why medicine, without being a science itself, uses the results of all the sciences in the service of the norms of life.

Thus it is first and foremost because men feel sick that a medicine exists. It is only secondarily that men know, because medicine exists, in what way they are sick.

Every empirical concept of disease preserves a relation to the axiological concept of disease. Consequently it is not an objective method which qualifies a considered biological phenomenon as pathological. It is always the relation to the individual patient through the intermediary of clinical practice, which justifies the qualification of pathological. While admitting the importance of objective methods of observation and analysis in pathology, it does not seem possible that we can speak with any correct logic of "objective pathology." Certainly a pathology can be methodical, critical and fortified experimentally. It can be called objective with reference to the physician who practices it. But the pathologist's intention is not that his object be a matter without subjectivity. One can carry out objectively, that is impartially, research whose object cannot be conceived and constructed without being related to a positive and negative qualification, whose object is not so much a fact as a value.

SECTION TWO

NEW REFLECTIONS ON THE NORMAL AND THE PATHOLOGICAL (1963–1966)

Twenty Years Later...

In 1943 as teacher [*chargé d'enseignement*] in the Faculty of Letters at Strasbourg in Clermont Ferrand, I gave a course on *Norms and the Normal*. At the same time I was writing my doctoral thesis in medicine, which I defended in July of the same year, before the Strasbourg Faculty of Medicine. In 1963 as professor in the Faculty of Letters and Social Sciences at Paris I gave a course on the same subject: twenty years later I wanted to measure myself against the same difficulties by other means.

It was out of the question simply to reexamine the same questions. Certain propositions, which I tried to support soundly in my *Essay* because of their – perhaps only apparent – paradoxical character seemed to me after that to be taken for granted: less because of the force of my own argumentation than because of the ingenuity of some readers who were clever in finding antecedents unknown to me. One young colleague,[1] a fine Kant specialist studying the Kantian philosophy in its relations with eighteenth-century biology and medicine, pointed out a text to me of the kind that generates at once the satisfaction of a great meeting and the embarrassment at an ignorance under whose shelter one believed one was able to claim for oneself a bit of originality. Kant noted, more than likely around 1798:

> The need to unravel the skein of politics by starting from the subjects' duties rather than the citizens' rights has recently been stressed. Likewise it is diseases which have stimulated physiology; and it is not physiology but pathology and clinical practice which gave medicine its start. The reason is that as a matter of fact well-being is not felt, for it is the simple awareness of living, and only its impediment provokes the force of resistance. It is no wonder then that Brown begins by classifying diseases. [Kant, *Werke*, Akademie Ausgabe, 15², *Anthropologie*, in the "Handschriftlicher Nachlass," p. 964].

Because of this it seemed superfluous to look for new justifications for the thesis which presents clinical practice and pathology as the breeding ground in which physiology is rooted, and as the path on which the human experience of disease conveys the concept of normal right to the heart of the physiologist's problematic. To this was added the fact that new readings of Claude Bernard, stimulated and clarified by the 1947 publication of the *Principes de médecine expérimentale*, necessarily softened the rigor of the judgment I first passed on his idea of the relations between physiology and pathology,[2] and made me more sensitive to the fact that Bernard did not ignore the need for clinical experience to precede laboratory experimentation.

> If I had to deal with beginners, the first thing I would tell them is go to the hospital; that is the first thing to get acquainted with. For how would one analyze diseases, which one didn't know, by means of experimentation? Therefore I am not saying substitute the laboratory for the hospital. I am saying the opposite: go to the hospital first, but this is not enough to attain scientific or experimental medicine; we must go to the laboratory afterwards to *analyze* experimentally what clinical

> observation has led us to record. I cannot imagine why this objection is made to me for I have indeed often said and repeated that medicine must always begin with a clinical observation (see *Introduction*, p. 242) and it is in this way that it began in ancient times.[3]

Conversely, having given Bernard his due, which I had in part contested, I had to show myself, as I also did, rather less generous with regard to Leriche.[4]

For all these reasons, my 1963 course explored the subject by tracing different paths from those of 1943. Other reading stimulated my reflections in other ways. It is not just a matter of reading works which have appeared in the interim. It is also a matter of readings which I could have or had done at the time. The bibliography of a subject always has to be redone, even retrospectively. One will understand this by comparing even here the 1966 bibliography with that of 1943.

But the two courses on *Norms and the Normal* by extension went beyond the subject of medical philosophy dealt with by the *Essay* which I still intend to reexamine in the pages that follow. The meaning of the concepts of norm and normal in the social sciences, sociology, ethnology, economics, involve research which in the end – whether it deals with social types, criteria of maladjustment to the group, consumer needs and behavior, preference systems – tends toward the question of the relations between normality and generality. If at the start I borrow some elements of analysis from the lectures, in which I examined some aspects of this question in my own way, it is only to clarify the specific meaning of vital norms by comparing them with social norms. It is with the organism in view that I am allowing myself some forays into society.

Can I confess that reading studies written after my 1943 thesis with a similar objective has not convinced me that I myself posed the problem badly at that time? All those who, like me, aimed at determining the meaning of the concept of normal have experienced the same difficulty and, faced with the term's polysemous character, had no other resource than to determine decisively the meaning which seemed to them most adequate for the theoretical or practical project which called up a semantic delimitation. This amounts to saying that those who themselves tried most rigorously to give "normal" only the value of a fact have simply valorized the fact of their need for a limited meaning. Today then, as twenty years ago, I am still running the risk of trying to establish the fundamental meaning of the normal by means of a philosophical analysis of life understood as activity of opposition to inertia and indifference. Life tries to win against death in all the senses of the verb to win, foremost in the sense of winning in gambling. Life gambles against growing entropy.

Chapter I

From the Social to the Vital

In the *Critique of Pure Reason* ([in the 3rd section of the] transcendental methodology: architectonic of pure reason), Kant distinguishes concepts, according to their sphere of origin and validity, into *scholastic* and *cosmic*, the latter serving as the foundation for the former.

We could say of the two concepts of Norm and Normal that the first is scholastic while the second is cosmic or popular. It is possible for the normal to be a category of popular judgment because their social situation is keenly, though confusedly, felt by the people as not being in line, not "right" (*droite*). But the very term normal has passed into popular language and has been naturalized there starting with the specific vocabularies of two institutions, the pedagogical institution and the hospital whose reforms, at least in France, coincided under the effect of the same cause, the French Revolution. "Normal" is the term used by the nineteenth century to designate the scholastic prototype and the state of organic health. The reform of medicine as theory, itself rests on the reform of medicine as practice: in France as also in Austria it is closely tied to hospital reform. Like pedagogical reform, hospital reform expresses a demand for rationalization which also appears in politics, as it appears in the economy, under the effect of nascent industrial mechanization, and which finally ends up in what has since been called normalization.

Just as a normal school is a school where teaching is taught, that is, where pedagogical methods are set up experimentally, so a normal medicine dropper is one which is calibrated to divide one gram of distilled water into twenty free-flowing drops so that the pharmaco-dynamic power of a substance in solution can be graduated according to a medical prescription. Again, among the 21 railway gauges used long ago and not so long ago, a normal track is one defined by the 1.44 meter gauge between the insides of the rails, that is, that track which, at a given moment of European industrial and economic history, seemed to correspond to the best compromise sought among several initially conflicting requirements related to mechanics, fuel, trade, the military and politics. Likewise, for the physiologist, man's normal weight, bearing in mind sex, age and height, is the weight "corresponding to the greatest predictable longevity."[5]

In the first three of these examples, the normal seems to be the effect of a choice and a decision external to the object so qualified, while in the fourth, the term of reference and qualification clearly appears as intrinsic to the object, if it is true that an individual organism's life span is a specific constant where health is maintained.

But when we think about it carefully, the normalization of the technical means of education, health, transportation for people and goods, expresses collective demands which, taken as a whole, even in the absence of an act of awareness [*prise de conscience*] on the part of individuals, in a given historical society, defines its way of referring its structure, or perhaps its structures, to what it considers its own good.

In any case the property of an object or fact, called normal in reference to an external or immanent norm, is the ability to be considered, in its turn, as the reference for objects or facts which have yet to be in a position to be called such. The normal is then at once the extension and the exhibition of the norm. It increases the rule at the same time that it points it out. It asks for everything outside, beside and against it that still escapes it. A norm draws its meaning, function and value from the fact of the existence, outside itself, of what does not meet the requirement it serves.

The normal is not a static or peaceful, but a dynamic and polemical concept. Gaston Bachelard, who was very preoccupied with values in their cosmic or popular form and in valorization following the axes of the imagination, has rightly perceived that every value must be earned against an anti-value. It is he who writes: "The will to cleanse requires an adversary its size."[6] When we know that *norma* is the Latin word for T-square and that *normalis* means perpendicular, we know almost all that must be known about the area in which the meaning of the terms "norm" and "normal" originated, which have been taken into a great variety of other areas. A norm, or rule, is what can be used to right, to square, to straighten. To set a norm (*normer*), to normalize, is to impose a requirement on an existence, a given whose variety, disparity, with regard to the requirement, present themselves as a hostile, even more than an unknown, indeterminant. It is, in effect, a polemical concept which negatively qualifies the sector of the given which does not enter into its extension while it depends on its comprehension. The concept of right, depending on whether it is a matter of geometry, morality or technology, qualifies what offers resistance to its application of twisted, crooked or awkward.[7]

The reason for the polemical final purpose and usage of the concept of norm must be sought, as far as we are concerned, in the essence of the normal–abnormal relationship. It is not a question of a relationship of contradiction and externality but one of inversion and polarity. The norm, by devaluing everything that the reference to it prohibits from being considered normal, creates on its own the possibility of an inversion of terms. A norm offers itself as a possible mode of unifying diversity, resolving a difference, settling a disagreement. But to offer oneself is not to impose oneself. Unlike a law of nature, a norm does not necessitate its effect. That is to say, a norm has no significance as norm pure and simple. Because we are dealing with possibility only, that possibility of reference and regulation which the norm offers leaves room for another possibility, which can only be its opposite. A norm is in effect the possibility of a reference only when it has been established or chosen as the expression of a preference and as the instrument of a will to substitute a satisfying state of affairs for a disappointing one. Every preference for a possible order is accompanied, most often implicitly, by the aversion for the opposite possible order. That which diverges from the preferable in a given area of evaluation is not the indifferent but the repulsive or more exactly, the repulsed, the detestable. It is well understood that a gastronomical norm does not enter into a relation of axiological opposition with a logical norm. On the other hand, the logical norm in which the true prevails over the false can be inverted into a norm where the false prevails over the true, as the ethical norm, where sincerity prevails over duplicity, can be inverted into a norm where duplicity prevails over sincerity. Yet the inversion of a logical norm does not yield a logical, but perhaps an aesthetic norm, as the inversion of an ethical norm does not yield an ethical, but perhaps a political one. In short, norms, whether in some implicit or explicit form, refer the real to values, express discriminations of qualities in conformity with the polar opposition of a positive and a negative. This polarity of the experience of normalization, a specifically anthropological or cultural experience – if it is true that by nature, only an ideal of normality without normalization must be understood – builds into the relationship of the norm to its area of application the normal priority of infraction.

In anthropological experience a norm cannot be original. Rule begins to be rule only in making rules and this function of correction arises from infraction itself. A golden age, a paradise, are the mythical representations of an existence which initially meets its demands, of a mode of life whose regularity owes nothing to the establishment of rules, of a state of guiltlessness in the absence of the interdict that ignorance of the law is no excuse. These two myths proceed from an illusion of retroactivity according to which original good is later evil kept in control. The absence of rules goes hand in hand with the absence of technical skills. Golden age man, and paradisiacal man, spontaneously enjoy the fruits of a nature which is uncultivated, unprompted, unforced, unreclaimed. Neither work nor culture, such is the desire of complete regression. This formulation in negative terms of an experience consonant with the norm without the norm having had to show itself in and by its function, this really naive dream of regularity in the absence of rule, signifies essentially that the concept of normal is itself normative, it serves as a norm even for the universe of mythical discourse which tells the story of its absence. This explains why, in many mythologies, the advent of the golden age marks the end of a chaos. As Gaston Bachelard said: "*Multiplicity is agitation*. In literature there is not one *immobile chaos*" [*op. cit.*, p. 59]. In Ovid's *Metamorphoses* the earth of chaos does not bear fruit, the sea of chaos is not navigable, forms do not remain identical to themselves. The initial indetermination is later denied determination. The instability of things has as its correlative the impotence of man. The image of chaos is that of a denied regularity, as that of the golden age is that of wild [*sauvage*] regularity. Chaos and golden age are the mythical terms of the fundamental normative relation, terms so related that neither of the two can keep from turning into the other. The role of chaos is to summon up, to provoke its interruption and to become an order. Inversely, the order of the golden age cannot last because wild regularity is mediocrity; the satisfactions there are modest – *aurea mediocritas* – because they are not a victory gained over the obstacle of measure. Where a rule is obeyed without awareness of a possible transcendence, all enjoyment is simple. But can one simply enjoy the value of rule itself? In order to truly enjoy the value of the rule, the value of regulation, the value of valorization, the rule must be subjected to the test of dispute. It is not just the exception which proves the rule as rule, it is the infraction which provides it with the occasion to be rule by making rules. In this sense the infraction is not the origin of the rule but the origin of regulation. It is in the nature of the normative that its beginning lies in its infraction. To use a Kantian expression, we would propose that the condition of the possibility of rules is but one with the condition of the possibility of the experience of rules. In a situation of irregularity, the experience of rules puts the regulatory function of rules to the test.

What eighteenth-century philosophers called the state of nature is the supposedly rational equivalent of the golden age. We must recognize with Lévi-Strauss that Rousseau, unlike Diderot, never thought that the state of nature was a historical origin for humanity brought to the ethnographer's attention by the geographer's exploration.[8] For his part[9] Jean Starobinski has shown successfully that the state of nature described by Rousseau is the portrayal of spontaneous equilibrium between the world and the values of desire, a state of prehistoric haphazardness in the absolute sense of the term, since it is from its irremediable disintegration that history flows as from a source. Strictly speaking, then, there is no grammatical tense adequate for a discussion of a human experience which has been normalized without the representation, in the consciousness, of norms linked to the temptation to oppose their exercise. For, either the adequation of fact and law is unperceived and the state of nature is a state of unawareness of which no event can explain that from it stems the occasion of a grasp of consciousness; or, the adequation is perceived and the state of nature is a state of innocence. But this state cannot exist for itself and be a state at the same time, that is, a static disposition. No one innocently knows that he is innocent since being aware of adequation to the rule means being aware of the reasons for the rule which amounts to the need for the rule. It is appropriate to contrast to the overly exploited Socratic maxim that no knowing man is evil, the opposite maxim that no one is good who is aware of being so. Similarly no one is healthy who knows that he is so. Kant's words: "Well-being is not felt for it is the simple consciousness of living"[10] are echoed by Leriche's definition: "Health is life in the silence of the organs." But it is in the rage of guilt as in the clamor of suffering that innocence and health arise as the terms of a regression as impossible as it is sought after.

The abnormal, as ab-normal, comes after the definition of the normal, it is its logical negation. However, it is the historical anteriority of the future abnormal which gives rise to a normative intention. The normal is the effect obtained by the execution of the normative project, it is the norm exhibited in the fact. In the relationship of the fact there is then a relationship of exclusion between the normal and the abnormal. But this negation is subordinated to the operation of negation, to the correction summoned up by the abnormality. Consequently it is not paradoxical to say that the abnormal, while logically second, is existentially first.

The Latin word *norma* which, etymologically speaking, bears the weight of the initial meaning of the terms "norms" and "normal," is the equivalent of the Greek ὀρθος. Orthography [French, *orthographe,* but long ago *orthographie*], orthodoxy, orthopedics, are normative concepts prematurely. If the concept of orthology is less familiar, at least it is not altogether useless to know that Plato guaranteed it[11] and the word is found, without a reference citation, in Littré's *Dictionnaire de la langue française*. Orthology is grammar in the sense given it by Latin and medieval writers, that is, the regulation of language usage.

If it is true that the experience of normalization is a specifically anthropological or cultural experience, it can seem normal that language has proposed one of its prime fields for this experience. Grammar furnishes prime material for reflection on norms. When Francis I in the edict of Villers-Cotterêt ordains that all judicial acts of the kingdom be drawn up in French, we are dealing with an imperative.[12] But a norm is not an imperative to do something under pain of juridical sanctions. When the grammarians of the same era undertook to fix the usage of the French language, it was a question of norms, of determining the reference and of defining mistakes in terms of divergence, difference. The reference is borrowed from usage. In the middle of the seventeenth century this is Vaugelas's thesis: "Usage is that to which we must subject ourselves entirely in our language."[13] Vaugelas's works turn up in the wake of works of the *Académie française* which was founded precisely to embellish the language. In fact in the seventeenth century the grammatical norm was the usage of cultured, bourgeois Parisians, so that this norm reflects a political norm: administrative centralization for the benefit of royal power. In terms of normalization there is no difference between the birth of grammar in France in the seventeenth century and the establishment of the metric system at the end of the eighteenth. Richelieu, the members of the National Convention and Napoleon Bonaparte are the successive instruments of the same collective demand. It began with grammatical norms and ended with morphological norms of men and horses for national defense,[14] passing through industrial and sanitary norms.

Defining industrial norms assumes a unity of plan, direction of work, stated purpose of material constructed. The article on "Gun-carriage" in the *Encyclopédie* of Diderot and d'Alembert, revised by the Royal Artillery Corps, admirably sets forth the motifs of the normalization of work in arsenals. In it we see how the confusion of efforts, the detail of proportions, the difficulty and slowness of replacements, useless expense, are remedied. The standardization of designs of pieces and dimension tables, the imposition of patterns and models have as their consequence the precision of separate products and the regularity of assembly. The "Gun-carriage" article contains almost all the concepts used in a modern treatise on normalization except the term norm. Here we have the thing without the word.

The definition of sanitary norms assumes that, from the political point of view, attention is paid to populations' health considered statistically, to the healthiness of conditions of existence, and to the uniform dissemination of preventive and curative treatments perfected by medicine. In Austria Maria Theresa and Joseph II conferred legal status on public health institutions by creating an Imperial Health Commission (*Sanitäts-Hofdeputation,* 1753) and by promulgating a *Haupt Medizinal Ordnung*, replaced in 1770 by the *Sanitäts-normativ*, an act with 40 regulations related to medicine, veterinary art, pharmacy, the training of surgeons, demographical and medical statistics. With respect to norm and normalization here we have the word with the thing.

In both of these examples the norm is what determines the normal starting from a normative decision. As we are going to see, such a decision regarding this or that norm is understood only within the context of other norms. At a given moment the experience of normalization cannot be broken down, at least not into projects. Pierre Guiraud clearly perceived this in the case of grammar when he wrote:

> Richelieu's founding of the *Académie française* in 1635 fit into a general policy of centralization of which the Revolution, the Empire, and the Republic are the heirs.... It would not be absurd to think that the bourgeoisie annexed the language at the same time that it seized the instruments of production.[15]

It could be said in another way by trying to substitute an equivalent for the Marxist concept of the ascending class. Between 1759, when the word "normal" appeared, and 1834 when the word "normalized" appeared, a normative class had won the power to identify – a beautiful example of ideological illusion – the function of social norms, whose content it determined, with the use that that class made of them.

That the normative intention in a given society in a given era cannot be broken down is apparent when we examine the relations between technological and juridical norms. In the rigorous and present meaning of the term, technological normalization consists in the choice and determination of material, the form and dimensions of an object whose characteristics from then on become necessary for consistent manufacture. The division of labor constrains businessmen to a homogeneity of norms at the heart of a technical–economic complex whose dimensions are constantly evolving on a national or international scale. But technology develops within a society's economy. A demand to simplify can appear urgent from the technological point of view but it can seem premature from the industrial and economic point of view as far as the possibilities of the moment and the immediate future are concerned. The logic of technology and the interests of the economy must come to terms. Moreover, in another respect, technological normalization must beware of an excess of rigidity. What is manufactured must finally be consumed. Certainly the logic of normalization can be pushed as far as the normalization of needs by means of the persuasion of advertising. For all that, should the question be settled as to whether need is an object of possible normalization or the subject obliged to invent norms? Assuming that the first of these two propositions is true, normalization must provide for needs, as it does for objects characterized by norms, margins for divergence, but here without quantification. The relation of technology to consumption introduces into the unification of methods, models, procedures and proofs of qualification, a relative flexibility, evoked furthermore by the term "normalization," which was preferred in France in 1930 to "standardization," to designate the administrative organism responsible for enterprise on a national scale.[16] The concept of normalization excludes that of immutability, includes the anticipation of a possible flexibility. So we see how a technological norm gradually reflects an idea of society and its hierarchy of values, how a decision to normalize assumes the representation of a possible whole of correlative, complementary or compensatory decisions. This whole must be finished in advance, finished if not closed. The representation of this totality of reciprocally relative norms is planning. Strictly speaking, the unity of a Plan would be the unity of a unique thought. A bureaucratic and technocratic myth, the Plan is the modern dress of the idea of Providence. As it is very clear that a meeting of delegates and a gathering of machines are hard put to achieve a unity of thought, it must be admitted that we would hesitate to say of the Plan what La Fontaine said of Providence, that it knows what we need better than we do.[17] Nevertheless – and without ignoring the fact that it has been possible to present normalization and planning as closely connected to a war economy or the economy of totalitarian regimes – we must see above all in planning endeavors the attempts to constitute organs through which a society could estimate, foresee and assume its needs instead of being reduced to recording and stating them in terms of accounts and balance sheets. So that what is denounced, under the name of rationalization – the bogey complacently waved by the champions of liberalism, the economic variety of the cult of nature – as a mechanization of social life, perhaps expresses, on the contrary, the need, obscurely felt by society, to become the organic subject of needs recognized as such.

It is easy to understand how technological activity and its normalization, in terms of their relation to the economy, are related to the juridical order. A law of industrial property, juridical protection of patents or registered patterns, exists. To normalize a registered pattern is to proceed to industrial expropriation. The requirement of national defense is the reason invoked by many States to introduce such provisions into legislation. The universe of technological norms opens onto the universe of juridical norms. An expropriation is carried out according to the norms of law. The magistrates who decide, the bailiffs responsible for carrying out the sentence, are persons identified with their function by virtue of norms, installed in their function with the delegation of competence. Here the normal descends from a higher norm through hierarchized delegation. In his *Reine Rechtslehre* (Leipzig, F. Deuticke, 1934, 2nd revised and enlarged edition, 1960; translated as *Pure Theory of Law*, 2nd revised and enlarged edition, Berkeley, University of California Press, 1967), Kelsen maintains that the validity of a juridical norm depends on its insertion in a coherent system, an order of hierarchized norms, drawing their binding power from their direct or indirect reference to a fundamental norm. But there are different juridical orders because there are several fundamental, irreducible norms. If it has been possible to contrast this philosophy of law with its powerlessness to absorb political fact into juridical fact, as it claims to do, at least its merit in having brought to light the relativity of juridical norms hierarchized in a coherent order has been generally recognized. So that one of Kelsen's most resolute critics can write: "The law is the system of conventions and norms destined to orient all behavior inside a group in a well-defined manner."[18] Even while recognizing that the law, private as well as public, has no source other than a political one, we can admit that the opportunity to legislate is given to the legislative power by a multiplicity of customs which must be institutionalized by that power into a virtual juridical whole. Even in the absence of the concept of juridical order, dear to Kelsen, the relativity of juridical norms can be justified. This relativity can be more or less strict. There exists a tolerance for non-relativity which does not mean a gap in relativity. In fact the norm of norms remains convergence. How could it be otherwise if law "is only the regulation of social activity?"[19]

To sum up, starting with the deliberately chosen example of the most artificial normalization, technological normalization, we can grasp an invariable characteristic of normality. Norms are relative to each other in a system, at least potentially. Their co-relativity within a social system tends to make this system an organization, that is, a unity in itself, if not by itself and for itself. One philosopher, at least, has noticed and brought to light the organic character of moral norms, much as they are first of all social norms. It is Bergson in *Les deux sources de la morale et de la religion* ["The Two Sources of Morality and Religion"] analyzing what he calls "the totality of obligation."

The correlativity of social norms – technological, economic, juridical – tends to make their virtual unity an organization. It is not easy to say what the concept of organization is in relation to that of organism, whether we are dealing with a more general structure than the organism, both more formal and richer; or whether we are dealing with a model which, relative to the organism held as a basic type of structure, has been singularized by so many restrictive conditions that it could have no more consistency than a metaphor.

Let us state first that in a social organization, the rules for adjusting the parts into a collective which is more or less clear as to its own final purpose – be the parts individuals, groups or enterprises with a limited objective – are external to the adjusted multiple. Rules must be represented, learned, remembered, applied, while in a living organism the rules for adjusting the parts among themselves are immanent, presented without being represented, acting with neither deliberation nor calculation. Here there is no divergence, no distance, no delay between rule and regulation. The social order is a set of rules with which the servants or beneficiaries, in any case, the leaders, must be concerned. The order of life is made of a set of rules lived without problems.[20]

The inventor of the term and first concept of *sociology*, August Comte, in the lectures of the *Cours de philosophie positive*, which deal with what he then called social physics, did not hesitate to use the term "social organism" to designate society defined as a *consensus* of parts coordinated according to two relations, synergy and sympathy, concepts borrowed from the Hippocratic medical tradition. Organization, organism, system, *consensus* are used indifferently by Comte to designate the state of society.[21] As far back as that period, Comte distinguished between society and power, understanding the latter concept as the organ and regulator of spontaneous common action,[22] an organ distinct but not separate from the social body, a rational, artificial but not arbitrary organ of the "manifest spontaneous harmony which must always tend to rule between the whole and the parts of the social system."[23] Thus the relationship between society and government is itself a relationship of correlation, and the political order appears as the voluntary and artificial extension "of this natural and involuntary order toward which the various human societies necessarily and incessantly tend in any respect."[24]

We must wait for the *Système de politique positive* in order to see Comte limit the scope of the analogy he accepted in the *Cours* and to emphasize the differences which keep one from considering as equivalent the structure of an organism and the structure of a social organization. In the fifth chapter ("Théorie positive de l'organisme social") of the *Statique sociale* (1852), Comte insists on the fact that the composite nature of the collective organism differs profoundly from the indivisible constitution of the organism. Though functionally concurrent, the elements of the social body are capable of a separate existence. In this respect the social organism does contain some mechanistic characteristics. In the same respect, moreover, "the collective organism, because of its composite nature, possesses to a higher degree the important aptitude, which the individual organism shows only in a rudimentary state, namely the ability to acquire new, even essential organs."[25] Because of this, regulation, the integration of successively related parts into a whole, is a specific social need. To regulate the life of a society, family or city is to introduce into a society – at once more general and more noble because closer to the only concrete social reality – Humanity or Great-Being. Social regulation is religion and positive religion is philosophy, spiritual power, the general art of man's action on himself. This function of social regulation must have a distinct organ, the priest, whose temporal power is merely a subordinate means. Socially speaking, to regulate is to cause the spirit of the whole to prevail. So that the entire social organism, if it is smaller than the Great-Being, is regulated from without and from above. The regulator is subsequent to what it regulates: "In effect only preexisting powers can be regulated, except instances of metaphysical illusion where we believe we create them to the extent that we define them."[26]

We shall say otherwise – certainly not better, probably less well – namely that a society is both machine and organism. It would be only a machine if the collective's ends could not only be strictly planned but also executed in conformity with a program. In this respect certain contemporary societies with a socialist form of economy tend perhaps toward an automatic mode of functioning. But it must be acknowledged that this tendency still encounters obstacles in facts, and not just in the ill will of skeptical performers, which oblige the organizers to summon up their resources for improvisation. It can even be asked whether any society whatsoever is capable of both clearsightedness in determining its purposes and efficiency in utilizing its means. In any case the fact that one of the tasks of the entire social organization consists in its informing itself as to its possible purposes – with the exception of archaic and so-called primitive societies where purpose is furnished in rite and tradition just as the behavior of the animal organism is provided by an innate model – seems to show clearly that, strictly speaking, it has no intrinsic finality. In the case of society, regulation is a need in search of its organ and its norms of exercise.

On the other hand, in the case of the organism the fact of need expresses the existence of a regulatory apparatus. The need for food, energy, movement and rest requires, as a condition of its appearance in the form of anxiety and the act of searching, the reference of the organism, in a state of given fact, to an optimum state of functioning, determined in the form of a constant. An organic regulation of a homeostasis assures first of all the return to the constant when, because of variations in its relation to the environment, the organism diverges from it. Just as need has as its center the organism taken in its entirety, even though it manifests itself and is satisfied by means of one apparatus, so its regulation expresses the integration of parts within the whole though it operates by means of one nervous and endocrine system. This is the reason why, strictly speaking, there is no distance between organs within the organism, no externality of parts. The knowledge the anatomist gains from an organism is a kind of display in extensiveness. But the organism itself does not live in the spatial mode by which it is perceived. The life of a living being is, for each of its elements, the immediacy of the co-presence of all.

The phenomena of social organization are like a mimicry of vital organization in the sense that Aristotle says that art imitates nature. Here to imitate does not mean to copy but to tend to rediscover the sense of a production. Social organization is, above all, the invention of organs – organs to look for and receive information, organs to calculate and even make decisions. In the still rather summarily rational form that it takes in contemporary industrial societies, normalization summons up planning which, in its turn, requires the establishment of statistics of all kinds and their utilization through computers. Provided that it is possible to explain – other than metaphorically – the functioning of a circuit of cortical neurons using the functioning of an electronic analyzer in transistor form as a model, it is tempting, if not legitimate, today to attribute some, perhaps the less intellectual functions for which the human brain is the organ, to the computers in the technico-economic organizations they serve. As for the assimilation of social information by means of statistics being analogous to the assimilation of vital information by means of sense receptors, to our knowledge it is older. It was Gabriel Tarde, who, in 1890 in *Les lois de l'imitation*, was the first to attempt it.[27] According to him statistics is the summation of identical social elements. The spreading of its results tends to yield its contemporary "intelligence" about the social fact in the process of being realized. We can imagine, then, a statistical department and its role as a social sense organ although for the moment, says Tarde, it is only a kind of embryonic eye. It must be noted that the analogy proposed by Tarde rests on the conception that physiological psychology had at that time the function of a sense receptor, like the eye or ear, according to which sensible qualities such as color or sound synthesize the components of a stimulant into one specific unit which the physicist counts in a multiplicity of vibrations. So that Tarde could write that "our senses, each one separately and from its special point of view, makes our statistics of the external universe."

But the difference between the social machinery for receiving and elaborating information, on the one hand, and the living organ on the other, still persists in that the perfecting of both in the course of human history and the evolution of life, takes place according to inverse modes. The biological evolution of organisms has proceeded by means of stricter integration of organs and functions for contact with the environment and by means of a more autonomous internalization of the conditions of existence of the organism's components and the establishment of what Claude Bernard called the "internal environment." Whereas the historical evolution of human societies has consisted in the fact that collectivities less extensive than the species have multiplied and, as it were, spread their means of action in spatial externality and their institutions in administrative externality, adding machines to tools, stocks to reserves, archives to traditions. In society the solution to each new problem of information and regulation is sought in, if not obtained by, the creation of organisms or institutions "parallel" to those whose inadequacy, because of sclerosis and routine, shows up at a given moment. Society must always solve a problem without a solution, that of the convergence of parallel solutions. Faced with this, the living organism establishes itself precisely as the simple realization – if not in all simplicity – of such a convergence. As Leroi-Gourhan writes:

> From animal to man everything happens summarily as if brain were added to brain, each of the latest developed formations involving an increasingly subtle cohesion of all the earlier forms which continue to play their role.[28]

Inversely the same author shows that "all human evolution converges to place outside of man what in the rest of the animal world corresponds to specific adaptation,"[29] which amounts to saying that the externalization of the organs of technology is a uniquely human phenomenon.[30] It is not forbidden to consider the existence of a distance between social organs, that is, the collective technical means at man's disposal, as a specific characteristic of human society. It is to the extent that society is an externality of organs that man can dispose of it by representation and therefore by choice. So that to propose the model of the organism for human societies in search of more and more organization is essentially to dream of a return not even to archaic, but to animal, societies.

There is hardly need, therefore, to insist now on the fact that social organs, if they are reciprocally purpose and means in a social whole, do not exist through one another and through the whole by virtue of coordinating causalities. The externality of social machines in the organization is in itself no different from the externality of parts in a machine.

Social regulation tends toward organic regulation and mimics it without ceasing for all that to be composed mechanically. In order to identify the social composition with the social organism in the strict sense of the term, we should be able to speak of a society's needs and norms as one speaks of an organism's vital needs and norms, that is, unambiguously. The vital needs and norms of a lizard or a stickleback in their natural habitat are expressed in the very fact that these animals are very natural living beings in this habitat. But it is enough that one individual in any society question the needs and norms of this society and challenge them – a sign that these needs and norms are not those of the whole society – in order for us to understand to what extent social need is not immanent, to what extent the social norm is not internal, and finally, to what extent the society, seat of restrained dissent or latent antagonisms, is far from setting itself up as a whole. If the individual poses a question about the finality of the society, is this not the sign that the society is a poorly unified set of means, precisely lacking an end with which the collective activity permitted by the structure would identify? To support this we could invoke the analyses of ethnographers who are sensitive to the diversity of systems of cultural norms. Lévi-Strauss says:

> We then discover that no society is fundamentally good, but that none is absolutely bad; they all offer their members certain advantages, with the proviso that there is invariably a residue of evil, the amount of which seems to remain more or less constant and perhaps corresponds to a specific inertia in social life resistant to all attempts at organization.[31]

Chapter II

On Organic Norms in Man

As far as health and disease are concerned, and consequently as far as setting accidents right, correcting disorders or, as it is popularly said, remedying ills are concerned, there is a difference between an organism and a society, in that the therapist of their ills, in the case of the organism, knows in advance and without hesitation, what normal state to establish, while in the case of society, he does not know.

G.K. Chesterton, in a small book called *What's Wrong with the World*,[32] denounced the frequent tendency of political writers and reformers to determine the state of social ill before proposing its remedies, calling it "medical error." The quick, brilliant, ironic refutation of what he calls a sophism rests on this axiom:

> Because, though there may be doubt about the way in which the body broke down, there is no doubt at all about the shape in which it should be built up again . . . . Medical science is content with the normal human body and only seeks to restore it.[33]

If there is no hesitation about the finality of a medical treatment, this is not so, says Chesterton, when it comes to social problems.

For the determination of the ill assumes the prior definition of the normal social state and the search for this definition divides those who devote themselves to it.

> The social case is exactly the opposite of the medical case. We do not disagree, like doctors, about the precise nature of the illness, while agreeing about the nature of health.[34]

It is social welfare that is discussed in society, which means that what some consider a downright ill others seek out as health as a matter of course![35]

There is something serious in this humor. To say that "no doctor proposes to produce a new kind of man, with a new arrangement of eyes or limbs,"[36] is to recognize that an organism's norm of life is furnished by the organism itself, contained in its existence. And it is quite true that no doctor dreams of promising his patients anything more than a return to the state of vital satisfaction from which illness hurled them down.

But it happens that there is more humor in reality than in humorists. Just when Chesterton was praising doctors for accepting the fact that the organism provides them with the norm for their restorative activity, certain biologists began to conceive of the possibility of applying genetics to transform the norms of the human race. The first lectures of H. J. Muller (the geneticist famous for his experiments with induced mutations) to be concerned with contemporary man's social and moral obligation to interfere with himself in order to generally move himself up to a higher intellectual level, that is, in short, to vulgarize genius by means of eugenics, date from 1910. On the whole it is not a matter of an individual desire but of a social program whose initial fate would have seemed to Chesterton the most perfect confirmation of his paradox. In *Out of the Night*,[37] Muller proposed a collectivity without classes and without social inequalities as a social ideal to be realized, where techniques for preserving seminal fluid and for artificial insemination would allow women, whom a rational education had made proud to have such an honor, to bear and raise the children of men of genius, of Lenin or Darwin.[38] Now it is precisely in the Soviet Union, where the book was written, that Muller's manuscript, sent to high places where it was thought it would please, was judged severely and the Russian geneticist, who had acted as a go-between, fell in disgrace.[39] A social ideal based on a theory of heredity like genetics, which establishes the fact of human inequality by creating techniques to correct it, would not be welcome in a classless society.

Without forgetting that genetics offers biologists precisely the possibility of conceiving and applying a formal biology and consequently of transcending life's empirical forms by creating experimental living beings following other norms, we shall agree that up until now a human organism's norm is its coincidence with the organism itself, while we wait for the day when it will coincide with the calculations of a eugenic geneticist.

If social norms could be perceived as clearly as organic norms, men would be mad not to conform to them. As men are not mad and as there are no Wise Men, social norms are to be invented and not observed. The concept of wisdom was a concept filled with meaning for Greek philosophers because they conceived of society as a reality of an organic type, having an intrinsic norm, its own health, rules of measure, equilibrium and compensation, a replication and imitation, on the human scale, of the universal law which made a *cosmos* of the totality of beings. A contemporary biologist, W.B. Cannon, echoed the assimilation of juridical concepts with medical concepts in archaic Greek thought when he entitled the work in which he expounds the theory of organic regulations – homeostasis – *The Wisdom of the Body*.[40] To speak of the wisdom of the body leads one to understand that the living body is in a permanent state of controlled equilibrium, of disequilibrium which is resisted as soon as it begins, of stability maintained against disturbing influences originating without: it means, in short, that organic life is an order of precarious and threatened functions which are constantly reestablished by a system of regulations. In ascribing a wisdom to the body, Starling and Cannon repatriated to physiology a concept which medicine had once exported to politics. Yet Cannon, in his turn, could not help expanding the concept of homeostasis so that he gave it the power to clarify social phenomena, entitling his last chapter: "Relations of Biological and Social Homeostasis." But the analysis of these relations is a tissue of commonplaces of liberal sociology and parliamentary politics concerning the alternation – in which Cannon sees the effect of a compensation apparatus – between conservatism and reformism. As if this alternation, far from being the effect of an apparatus which is inherent, even in the rudimentary state, to every social structure, were not in fact the expression of the relative efficiency of a regime invented to channel and smother social antagonisms, of a political machine acquired by modern societies in order to delay, without finally being able to prevent, the transformation of their inconsistencies into crisis. In observing industrial age societies it can be asked whether their actual permanent state is not one of crisis and whether this is not an unequivocal symptom of the absence of their power of self-regulation.

The regulations for which Cannon invented the general term *homeostasis*[41] are similar to those which Claude Bernard had unified under the name of "constants of the internal environment." These are norms of organic functioning such as the regulation of respiratory movements under the effect of the rate of carbonic acid dissolved in the blood, thermoregulation in animals with constant temperature, etc. We know today what Bernard could only suspect, namely that other forms of regulation must be taken into consideration in studying organic structures and the origin of these structures. Contemporary experimental embryology has found its basic problems in the fact of morphological regulations which, in the course of embryonic development, conserve or establish the integrity of the specific form and extend their organizing action in repairing certain mutilations. So that the set of norms, by virtue of which living beings show themselves as forming a distinct world, can be classed as norms of constitution, norms of reconstitution and norms of functioning.

These different norms pose the same problem for biologists, namely their relation to uncommon cases which, in terms of the normal specific characteristic, show up a distance or a divergence of this or that biological characteristic: height, structure of an organ, chemical composition, behavior, etc. If the individual organism is the one which, of its own accord, proposes the norm for its restoration, in the case of malformation or accident, what sets up as norms the specific structure and functions which cannot be grasped by the individuals other than as they are manifested? Thermoregulation varies from the rabbit to the stork, from the horse to the camel. But how do we understand the norms peculiar to each species, rabbits, for example, without erasing the slight, fragmentary dissimilarities which give individuals their singularity?

The concept of *normal* in biology is objectively defined in terms of the frequency of the characteristic so qualified. For a given species, weight, height, maturation of instincts for a given age and sex are those which effectively characterize the most numerous groups distinctively formed by individuals of a natural population made to appear identical by measurement. It was Quetelet who observed around 1843 that the distribution of human heights could be represented by the error law established by Gauss, a limiting form of the binomial law, and who distinguished the two concepts of the Gaussian average or true average from the arithmetic average, which were at first identified in the theory of the average man. The distribution of the results of measurement on either side of the average value guarantees that the Gaussian average is the true average. The greater the divergences the rarer they are.

In our *Essay* (Part Two, II) we tried to preserve in the concept of norm a meaning analogous to that of the concept of type which Quetelet had superimposed on his theory of the average man following the discovery of the true average. It is an analogous meaning, that is, similar in function but different in foundation. Quetelet defined the regularity expressed by the average, by the greatest statistical frequency, as the effect of living beings' submission to laws of divine origin. We had tried to show that the frequency can be explained in terms of regulations of an order completely different from conformity to supernatural legislations. We had interpreted frequency as the actual or virtual criterion of the vitality of an adaptive solution.[42] We have to believe that our attempt missed its goal since it has been criticized for obscurity and for drawing the unwarranted conclusion that the greatest frequency equals the best adaptation.[43] In fact there is adaptation and adaptation, and the sense in which it is understood in the objections made to our work is not the sense we had given it. There is one form of adaptation which is specialization for a given task in a stable environment, but which is threatened by any accident which modifies this environment. And there is another form of adaptation which signifies independence from the constraints of a stable environment and consequently the ability to overcome the difficulties of living which result from a change in the environment. Now, we have defined a species' normality in terms of a certain tendency toward variety, "a kind of insurance against excessive specialization without reversibility, hence without flexibility, which is . . . a successful adaptation." In adaptation perfect or completed means the beginning of the end of species. At that time we were inspired by an article of the biologist Albert Vandel, who later developed the same ideas in his book, *L'homme et l'evolution.*[44] May we now be allowed to resume our analysis.

When the normal is defined in terms of the most frequent, a considerable obstacle is created to understanding the biological significance of those anomalies which geneticists have given the name of mutations. Indeed, to the extent to which a mutation in the plant or animal world can be the origin of a new species, we can see one norm arise from a divergence from another norm. The norm is the form of divergence maintained by natural selection. It is what destruction and death concede at random. But we know well that mutations are more often restrictive than constructive, often superficial when they are lasting, and when they are considerable, they involve fragility, a decrease in organic resistance. So that one acknowledges the power of mutations to diversify species into varieties rather than to explain the origin of species.

Strictly speaking, a mutationist theory of the origin of species can define the normal only as the temporarily viable. But by considering the living only as the dead with a respite, we ignore the adaptive orientation of the whole of living beings considered in the continuity of life, and we underestimate this aspect of evolution which is variation of modes of life for the occupation of all vacant places.[45] There is then one meaning of adaptation which allows one to distinguish, at a given moment in a species and its mutants, between obsolete and progressive living beings. Animality is a form of life characterized by mobility and predation. In this regard vision is a function which might not be called useless for mobility in light. A blind and cave-dwelling animal species can be said to be adapted to the dark and we can imagine its appearance by means of mutation, starting from a sighted species, and its maintenance by encountering and occupying an environment which, if not adequate, is at least not contra-indicated. Nonetheless, blindness is considered an anomaly, not in the sense that it is a rarity but in the sense that it means regression for the living beings concerned, a placing aside in a dead end.

It seems to us that one of the signs of the difficulty in explaining the specific norm in biology in terms of a single encounter of independent causal series, one biological, the other geographical, is the appearance, in 1954, of Lerner's concept of genetic homeostasis in population genetics.[46] The study of the arrangements of genes and the appearance of mutant genes in individuals in natural and experimental populations, combined with the study of the effects of natural selection, have led to the conclusion that the selective effect of a gene or of a certain arrangement of genes is not constant, that it undoubtedly depends on environmental conditions but also on a kind of pressure exerted on any one individual by the genetic totality represented by the population. Even in the case of human diseases, for example Cooley's anemia, which is common in the Mediterranean, particularly in Sicily and Sardinia, a selective superiority of heterozygote individuals over homozygotes has been observed. In animals on breeding farms this superiority can be measured experimentally. This coincides with the old observations of breeders concerning the invigoration of breeding lines by crossbreeding. Heterozygotes are more fertile. In the case of a lethal mutant gene, a heterozygote enjoys a selective advantage in relation not only to the mutant homozygote but also to the normal homozygote – whence the concept of genetic homeostasis. To the extent to which the survival of a population is favored by the frequency of heterozygotes, the proportional relation between fertility and heterozygosis can be considered a regulation. According to J.B.S. Haldane the same is true for a species' resistance to certain parasites. A biochemical mutation can obtain a greater capacity for resistance for the mutant. The individual biochemical difference at the heart of a species makes it more fit to survive, at the cost of alterations which morphologically and physiologically express the effects of natural selection. Unlike humanity which, according to Marx, poses only problems which it can solve, life multiplies beforehand the solutions to problems of adaptation which could present themselves.[47]

To summarize, the reading and reflecting which we have been able to do since the 1943 publication of our *Essay* have not led us to put in question the interpretation proposed then for the biological foundation of the original concepts of biometry.

It does not seem to us that we must profoundly modify our analysis of the relations between the determination of statistical norms and the evaluation of the normality or abnormality of this or that individual divergence. In the *Essay* we relied on the studies of André Mayer and Henri Laugier. Among the numerous articles published since on the same subject, two have claimed our attention.

The first belongs to A.C. Ivy: "What is Normal or Normality?" (1944).[48] The author distinguishes four meanings of the concept of normal: (1) coincidence between an organic fact and an ideal which decides the lower or upper limit of certain demands; (2) the presence in an individual of characteristics (structure, function, chemical composition) whose measure is conventionally determined by the central value of a group which is homogeneous in terms of age, sex, etc.; (3) an individual's situation in terms of the average for each characteristic considered, when the distribution curve has been constructed, the divergence type calculated and the number of divergence types determined; (4) the awareness of the absence of handicaps.

The use of the concept "normal" demands that one specify first the meaning by which one understands it. For his part, the author considers only numbers 3 and 4, subordinating the latter to the former. He applies himself to showing how desirable it is to establish the typical deviation of measurement values of structure, functions or biochemical components in a large number of subjects, especially when the results deviate strongly, and to consider as normal the values represented by 68.26% of an examined population, that is, the values corresponding to the average plus or minus a standard deviation. It is the subjects whose values fall outside the 68% who pose difficult evaluation problems in terms of their relation to the norm. For example: the temperature of 10,000 students, who are asked to say whether they feel feverish or not, is taken, the distribution of the temperatures is constructed, and for each group with the same temperature the correlation between the number of individuals and the number of subjects who say they are feverish, is calculated. The closer the correlation is to 1, the greater the chances are that the subject, because of an infection, is in a pathological state. Out of 50 subjects with a temperature of 100 F°, there is only a 14% chance for a normal subject from the subjective point of view (i.e., one who did not feel feverish) to be a normal subject from the bacteriological point of view.

The interest of Ivy's study lies less in the information from classical statistics than in the simplicity with which the author acknowledges the difficulties of the coincidence of concepts such as the physiological normal and the statistical normal. The state of physiological plenitude ("the healthful condition" [in Ivy's work]) is defined as a state of equilibrium of functions that are so integrated that they gain for the subject a large measure of security, a capacity for resistance in a critical situation or a situation of force. The normal state of a function is that of not interfering with others. But can it not be objected to these propositions that most functions, because of their integration, do interfere. If we must understand that a function is normal insofar as it does not lead another to abnormality, hasn't the question been shifted? In any case the comparison of these physiological concepts and the concept of norm statistically defined – the state of 68% of subjects in a homegeneous group – shows up the inability of the statistically defined norm to resolve a concrete problem of pathology. The fact that an old man exhibits functions included in the 68% corresponding to his age is not sufficient to qualify him as normal to the extent that the physiological normal is defined in terms of a margin of security in the exercise of functions. Aging is expressed, in effect, by the reduction of this margin. Finally an analysis such as Ivy's, starting from other examples, wants to confirm the inadequacy, often recognized before him, of the statistical point of view each time it must be decided as to what is normal or not for a given individual.

The necessity to rectify the concept of the statistical normal and to make it flexible in response to the physiologist's experience based on the variability of functions is also brought to light in an article of 1947 by John A. Ryle, "The Meaning of Normal."[49] The author, a professor of social medicine at Oxford, is interested first in establishing that certain individual divergences, in relation to physiological norms, are not, for all that, pathological indicators. It is normal for physiological variability to exist, it is necessary for adaptation and hence survival. The author examined 100 students in good health, free of dyspepsia, in whom he took measurements of gastric acidity. He ascertained that 10% showed what could be considered pathological hyperchlorhydria such as is observed in the case of duodenal ulcer, and that 4% showed total achlorhydria, a symptom considered until then as indicative of progressive pernicious anemia. The author thinks that all measurable physiological activities show themselves susceptible of an analogous variability, that they can be represented by the Gaussian curve and that, for the needs of medicine, the normal must be included between the limits determined by a standard deviation on both sides of the median. But there is no clear dividing line between innate variations compatible with health, and acquired variations which are the symptoms of a disease. If really necessary, one can think that an extreme physiological divergence in terms of the average constitutes or contributes to constituting a predisposition to this or that pathological accident.

John A. Ryle lists the medical activities, for which the concept of "normal clearly understood" corresponds to a need, as follows: (1) definition of the pathological; (2) definition of the functional levels to aim for in treatment or reeducation; (3) the choice of personnel employed in industry; (4) tracking down predispositions to disease.

Let us note, for it is not unimportant, that the last three needs of this list concern criteria of expertise, capacity, incapacity, mortality risk.

Finally Ryle distinguishes two kinds of variations relative to the norm, with regard to which it may be that one is to decide abnormality in view of certain resolutions to be taken of a practical order: variations affecting the same individual according to time; variations, at a given moment, from one individual to another in the species. These two kinds of variations are essential for survival. Adaptability depends on variability. But the study of adaptability must always be circumstantial, it is not enough to proceed to laboratory measurements and tests; the physical and social environment, diet, mode and conditions of work, the economic situation and education of different classes must also be studied, for as the normal is considered as the indicator of a fitness or an adaptability, we must always ask ourselves to what and for what must we determine adaptability and fitness. For example: the author reports the results of an investigation into thyroid enlargement in 11 to 15 year olds in areas where the amount of iodine in the drinking water has been precisely measured. In this case the normal is the thyroid which is externally inconspicuous. The conspicuous thyroid seems to indicate a specific mineral deficiency. But as few children with a conspicuous thyroid end up with goiter, it can be claimed that a clinically discernible hyperplasia expresses a degree of advanced adaptation rather than the first stage of disease. Since the thyroid is always smaller in Icelanders, and since, on the other hand, there are areas in China where 60% of the inhabitants have goiters, it seems that we can speak of national standards of normality. In short, in order to define the normal, we must refer to concepts of equilibrium and adaptability, and bear in mind the external environment, and the work which the organism or its parts must accomplish.

The study we have just summarized is interesting, without being methodologically intolerant, and ends by concluding that the preoccupations of expertise and evaluation prevail over those of measurement in the strict sense of the word.

In dealing with human norms we acknowledge that they are determined as an organism's possibilities for action in a social situation rather than as an organism's functions envisaged as a mechanism coupled with the physical environment. The form and functions of the human body are the expression not only of conditions imposed on life by the environment but also of socially adopted modes of living in the environment. In our *Essay* we took into account observations which allowed us to consider an interdependence between nature and culture to be probable in determining human organic norms, from the fact of the psychosomatic relation.[50] At the time our conclusions might have seemed rash. Today it seems to us that the development of studies in psychosomatic and psychosocial medicine, particularly in Anglo-Saxon countries, would tend to confirm them. A well-known specialist in social psychology, Otto Klineberg, in a study on tensions related to international understanding,[51] has pointed out the psychosomatic and psychosocial causes of varieties of reactions and disturbances involving apparently lasting modifications of organic constants. Chinese, Hindus and Filipinos exhibit an average systolic pressure 15 to 20 points lower than Americans'. But the average systolic blood pressure of Americans who have passed several years in China fell during that period from 118 to 109. Similarly it could be noted that during the period 1920–1930 hypertension in China was very rare. While finding it "simplistic in the extreme" Klineberg cites the remark of an American doctor, made about 1929:

> If we stay in China long enough we learn to accept things and our blood pressure falls. Chinese in America learn to protest and to not accept and their blood pressure mounts.

To assume that Mao Tse-Tung has changed all that is not being ironic, but simply applying the same method of interpreting psychosocial phenomena to other political and social data.

The concept of adaptation and that of psychosomatic relation to which its analysis leads, in the case of man, can be taken up again and reworked, so to speak, as a function of theories of pathology which differ as to their basic observations, but which converge in spirit. Relating physiological norms in man to show up cultural norms, is naturally extended by the study of specifically human pathogenic situations. In man, unlike in laboratory animals, the pathogenic stimuli or agents are never received by the organism as brute physical facts, but are lived by the consciousness as signs of tasks or tests.

Hans Selye – almost at the same time as Reilly in France – was one of the first to tackle the study of nonspecific pathological syndromes, characteristic reactions and behaviors in every disease considered at its onset, from the general fact of "feeling sick."[52] A nonspecific aggression (i.e. a brusque stimulation) provoked by any stimulus whatsoever – foreign body, purified hormone, traumatism, pain, repeated emotion, imposed fatigue, etc. – triggers off first an alarm reaction, also nonspecific, consisting essentially in the wholesale excitation of the sympathetic nerve which is accompanied by an adrenalin and noradrenalin secretion. In short the alarm puts the organism in a state of emergency, a state of indefinite parrying. This alarm reaction is followed by either a specific state of resistance, as if the organism had identified the nature of the aggression, was adapting its response to the attack and was reducing its initial susceptibility to the outrage; or, by a state of exhaustion when the intensity and ceaselessness of the aggression exceed reaction capacities. These are Selye's three moments of the general adaptation syndrome. Adaptation is thus considered as the physiological function par excellence. We propose to define it as organic impatience with the indiscreet interventions or provocations of the environment, be it cosmic (action of physico-chemical agents) or human (emotions). If by physiology we understand the science of the functions of normal man, it must be recognized that this science rests on the postulate that normal man is the man of nature. As one physiologist, Bacq, wrote: "Tranquility, laziness, psychic indifference are decisive trumps for the maintenance of normal physiology."[53] But perhaps human physiology is always more or less applied physiology, physiology of work, of sport, of leisure, of life at high altitudes, etc., that is, the biological study of man in cultural situations which generate varied aggressions.[54] In this sense we would find in Selye's theories confirmation of the fact that norms are recognized by their divergences.

Under the name of "adaptation diseases" must be understood all kinds of disorders of the function of resistance to disturbances, diseases of the function of resistance to harm. By this let us understand reactions which go beyond their goal, which run on their impetus and persevere until the aggression has stopped. Now is the time to say with F. Dagognet:

> The sick person creates disease by the very excess of his defence and by the importance of a reaction which protects less than it exhausts and upsets. The remedies which inhibit or stabilize take precedence over all those which stimulate, enchance, or sustain.[55]

It is not within our competence to decide whether Selye's observations and those of Reilly and his school are identical and whether the humoral mechanisms invoked by one and the neurovegetative mechanisms invoked by the others are complementary or not.[56] We consider only the convergence of these theses on the following point: the predominance of the notion of pathogenic syndrome over that of pathogenic agent, the subordination of the notion of lesion to that of the disturbance of functions. In a famous lecture, contemporary with the early investigations of Reilly and Selye, P. Abrami drew attention to the number and importance of functional disturbances, which, from the point of view of the clinical symptomatology of identical lesions, are sometimes capable of diversifying, and above all in the long run, sometimes capable of giving birth to organic lesions.[57]

By now, we are far from the wisdom of the body. In effect, one could suspect as much by comparing adaptation diseases with all the phenomena of anaphylaxia, allergy, that is to say, all the organism's phenomena of hyperreactivity against an aggression to which it is sensitized. In this case disease consists in the immoderacy of the organic response, in the outburst and stubbornness of the defense, as if the organism aimed badly, calculated badly. The term "error" came naturally to the minds of pathologists to designate a disturbance whose origin is to be sought in the physiological function itself and not in the external agent. In identifying histamine, Sir Henry Dale had considered it as a product of "organic autopharmacology." From then on, can a physiological phenomenon which ends up in what Bacq calls "The veritable suicide of the organism by means of toxic substances which it stocks in its own tissues," be called anything other than error?[58]

Chapter III

A New Concept in Pathology: Error

In our *Essay* we compared the ontological conception of disease, in which disease is portrayed as the qualitative opposite of health, with the positivist conception, which derives it quantitatively from the normal state. When disease is considered as an evil, therapy is given for a revalorization; when disease is considered as deficiency or excess, therapy consists in compensation. Against Bernard's conception of disease we set the existence of illnesses such as alkaptonuria, whose symptom can in no way be derived from the normal state and whose process – the incomplete metabolism of tyrosine – bears no quantitative relation to the normal process.[59] It must be acknowledged today that even then our argument could have been further solidified by being more broadly buttressed with examples, by considering albinism and cystinuria.

Since 1909 these metabolic diseases, because they block reactions at an intermediary stage, have been given the striking name of "inborn errors of metabolism,"[60] a term coined by Sir Archibald Garrod. Hereditary biochemical disturbances, nevertheless these genetic diseases cannot manifest themselves as early as birth, but rather in the course of time and should the occasion present itself, as for example, in the human organism's lack of a diastase (glucose-6-phosphate-dehydrogenase) which gives rise to no disturbance as long as the subject does not introduce beans into his diet or take quinine to combat malaria. For fifty years medicine had recognized only half a dozen of these diseases, and they could be considered rarities. This explains why the concept of inborn metabolic error was not a common concept in pathology at the time we undertook our medical studies. Today hereditary biochemical diseases number about one hundred. The identification and treatment of some of the particularly distressing ones such as phenylketonuria or phenylpyruvic imbecility have given grounds for great hopes in extending the genetic explanation of diseases. The etiology of sporadic or endemic diseases such as goiter is being revised in the light of research on biochemical anomalies of a genetic nature.[61] So we can imagine that, while the concept of inborn error of metabolism has not become a popular concept, strictly speaking, it is nevertheless a common one today. The terms anomaly, lesion, borrowed from the language of morphological pathology, have been imported into the domain of biochemical phenomena.[62]

At the outset, the concept of hereditary biochemical error rested on the ingenuity of a metaphor; today it is based on the solidity of an analogy. Insofar as the fundamental concepts of the biochemistry of amino acids and macromolecules are concepts borrowed from information theory, such as code or message; and insofar as the structures of the matter of life are linear structures, the negative of order is inversion, the negative of sequence is confusion, and the substitution of one arrangement for another is error. Health is genetic and enzymatic correction. To be sick is to have been made false, to be false, not in the sense of a false bank note or a false friend, but in the sense of a "false fold" [i.e., wrinkle: *faux pli*] or a false rhyme. Since enzymes are the mediators through which the genes direct intracellular protein syntheses, and since the information necessary for this function of direction and surveillance is inscribed in the DNA molecules at the chromosome level, this information must be transmitted as a message from the nucleus to the cytoplasm and must be interpreted there, so that the sequence of amino acids constituting the protein to be synthesized is reproduced, recopied. But whatever the mode, there is no interpretation which does not involve a possible mistake. The substitution of one amino acid for another creates disorder through misunderstanding the command. For example, in the case of sickle-cell anemia, that is, red blood cells shaped like a sickle because of retraction following a lowering of oxygen pressure, the hemoglobin is abnormal because of the substitution of valine for glutamic acid in the globulin's amino-acid chain.

The introduction of the concept of error into pathology is a fact of great importance as much in terms of the change it reveals in what it brings to bear in man's attitude toward disease, as in terms of the new status which is supposedly established in the relationship between knowledge and its object. It would be very tempting to denounce an identification of thought and nature, to protest that the steps of thought are ascribed to nature, that error is characteristic of judgment, that nature can be a witness, but never a judge, etc. Apparently everything happens, in effect, as if the biochemist and geneticist attributed their knowledge as chemist and geneticist to the elements of the hereditary patrimony, as if enzymes were supposed to know or must know the reactions according to which chemistry analyzes their action and could, in certain instances or at certain times, ignore one of them or misread the terms. But it must not be forgotten that information theory cannot be broken down, and that it concerns knowledge itself as well as its objects, matter or life. In this sense to know is to be informed, to learn to decipher or decode. There is then no difference between the error of life and the error of thought, between the errors of informing and informed information. The first furnishes the key to the second. From the philosophical point of view it would be a question of a new kind of Aristotelianism, on the condition, of course, that Aristotelian psychobiology and the modern technology of transmission not be confused.[63]

In certain respects this notion of error in the biochemical composition of this or that constituent of an organism is also Aristotelian. According to Aristotle, a monster is an error of nature which was mistaken about matter. If in contemporary molecular pathology, error generates formal flaws, hereditary biochemical errors are always considered as a microanomaly, a micromonstrosity. And just as a certain number of congenital morphological anomalies are interpreted as a fixation of the embryo at a stage of development which should normally be passed through, so a certain number of metabolic errors are like an interruption or suspension of a series of chemical reactions.

In such a conception of disease the harm is truly radical. If it manifests itself at the level of the organism taken as a whole, at grips with an environment, it remains at the very roots of the organization, at the level where it is still only linear structure, where not the domain but the order of the living being begins. Disease is not a fall that one has, an attack to which one succumbs, but an original flaw in macromolecular form. If, in principle, organization is a kind of language, the genetically determined disease is no longer a mischievous curse but a misunderstanding. There are bad readings of a hemoglobin just as there are bad readings of a manuscript. But here we are dealing with a word which comes from no mouth, with a writing which comes from no hand. There is then no ill will behind the ill fate. To be sick is to be bad, not as a bad boy but as poor land. Disease is no longer related to individual responsibility; no more imprudence, no more excess to incriminate, not even collective responsibility as in the case of epidemics. As living beings, we are the effect of the very laws of the multiplication of life, as sick men we are the effect of universal mixing, love and chance. All this makes us unique, as has often been written to console us for having been made from balls drawn by lot in the urn of Mendelian heredity. Unique, certainly, but sometimes also badly turned out. It is not too serious if it is only a matter of error in the metabolism of fructose because of the lack of hepatic aldolase.[64] It is more serious if it is a question of hemophilia, arising from the lack of synthesis of a globulin. And what is to be said, if not inadequately, if we are dealing with an error in the metabolism of tryptophane which, according to J. Lejeune, determines Mongolian trisomy?

The term error mobilizes the affectivity less than the terms disease or ill, wrongly nevertheless, if it is true that error is, at the outset, miscarriage. This is why the introduction of theoretical illusion into the vocabulary of pathology lets certain people hope, perhaps, for progress toward rationality in negative vital values. In fact when the eradication of error is obtained, it is irreversible, while the cure for a disease is sometimes the open door to another, hence the paradox of "diseases which are dangerous to cure."[65]

Nevertheless it can be maintained that the notion of innate organic errors is anything but reassuring. It takes a great deal of clarity coupled with great courage not to prefer an idea of disease where some feeling of individual culpability can still find a place in an explanation of disease which pulverizes and scatters its causality in the familial genome, in a legacy which the legatee cannot refuse since legacy and legatee are but one. But in the end it must be admitted that the notion of error, like the concept of pathology, is polysemic. If it consists, at the outset, in a confusion of formula, a false taken for the truth, it is recognized as such at the conclusion of research which has been stimulated by the difficulty of living, or by pain, or by someone's death. In relation to the denial of death, pain, difficulty in living, that is, to medicine's *rai-*

sons d'être, the error of enzymatic reading is experienced by the man who suffers from it as a fault of conduct without the fault of the conductor. In short, the use of the term designating the logical fault does not succeed in totally exorcizing from medical semantics the traces of anguish felt with the idea which we must count with an original abnormality.

Less reassuring still is the idea that it is appropriate to develop a medical counterpart to hereditary errors when this idea is formed as an idea and not as a desire. By definition a treatment cannot put an end to what is not the consequence of an accident. Heredity is the modern name of substance. We can imagine that it is possible to neutralize the effects of an error of metabolism by constantly furnishing the organism with the reaction product which is indispensable to the exercise of that function, of which an incomplete chain of reactions deprives it. And this is what is successfully done in the case of phenylpyruvic oligophrenia. But to compensate an organism's deficiencies for life only perpetuates a solution of distress. The real solution to heresy is extirpation. Consequently why not dream of hunting for heterodox genes, of a genetic inquisition? And while waiting, why not deprive suspect sires of the liberty of sowing broadcast? We know that these dreams are not only dreams for some biologists of very different philosophical persuasion, if we may call it that. But in dreaming these dreams, we enter another world, bordering on the bravest of Aldous Huxley's from which sick individuals, their particular diseases and their doctors have been eliminated. The life of a natural population is portrayed as a lotto sack and the functionaries delegated by the life sciences have the task of verifying the regularity of its numbers before the players are allowed to draw them from the sack to fill their cards. At the beginning of this dream we have the generous ambition to spare innocent and impotent living beings the atrocious burden of producing errors of life. At the end there are the gene police, clad in the geneticists' science. For all that it should not be concluded that one is obliged to respect a genetic "laisser-faire, laisser-passer," but only obliged to remind medical consciousness that to dream of absolute remedies is often to dream of remedies which are worse than the ill.

If diseases caused by innate chemical malformations are numerous as to their variety, none is widespread. If it were otherwise, the concept of the wisdom of the body would seem very irrelevant, to which one may answer that the errors of organization do not contradict the wisdom of organisms, that is, the results of organization. What was once true of finality is true today of organization. Against finality one has always invoked life's failures, the disharmony of organisms, or the rivalry of living species, macroscopic or microscopic. But if these facts represent objections to a real, ontological finality, they run counter to arguments supporting a possible, operative finality. If there were a perfect, finished finality, a complete system of relations of organic agreement, the very concept of finality would have no meaning as a concept, as a plan and model for thinking about life, for the simple reason that there would be no grounds for thought, no grounds for thinking in the absence of all disparity between possible organization and real organization. The thought of finality expresses the limitation of life's finality. If this concept has a meaning, it is because it is the concept of a meaning, the concept of a possible, and thus not guaranteed, organization.

In fact the explanation of the rarity of biochemical diseases lies in the fact that hereditary metabolic anomalies often remain hidden as nonactivated tendencies. In the absence of chance encounters with this component of life's environment or that effect of vital competition, these anomalies can be ignored by their bearers. Just as not all pathogenic germs determine an infection in any host under any circumstances, so not all biochemical lesions are someone's disease. In certain ecological contexts it even happens that they confer a certain superiority on those who must then be called their beneficiaries. In man, for example, a deficiency in glucose-6-phosphate-dehydrogenase has been diagnosed only when antimalarial drugs (quinine) were administered to Blacks in the United States. Now according to Dr. Henri Péquignot:

> When we study how an enzymatic disease, which is genetic, could maintain itself in the Black population, we realize that these subjects are in so much better shape that the "sick people" afflicted with this disturbance are particularly resistant to malaria. Their ancestors in black Africa were "normal" people in relation to others who were unfit, since they were resistant to malaria while the others died from it.[66]

While recognizing that certain innate biochemical errors receive their eventual pathological value from a relation between the organism and the environment, as certain lapses or mistaken acts, according to Freud, receive their value as symptoms from a relation to a situation, we are taking care not to define the normal and the pathological in terms of their simple relation to the phenomenon of adaptation. After a quarter of a century, this concept has received such an application in psychology and sociology, often inopportune, that it can only be used in the most critical spirit, even in biology. The psychosocial definition of the normal in terms of adaptedness implies a concept of society which surreptitiously and wrongly assimilates it to an environment, that is, to a system of determinisms when it is a system of constraints which, already and before all relations between it and the environment, contains collective norms for evaluating the quality of these relations. To define abnormality in terms of social maladaptation is more or less to accept the idea that the individual must subscribe to the fact of such a society, hence must accommodate himself to it as to a reality which is at the same time a good. Because of the conclusions of our first chapter, it seems legitimate to us to be able to refuse this kind of definition without being charged with anarchism. If societies are badly unified sets of means, they can be denied the right to define normality in terms of the attitude of instrumental subordination which they valorize under the name of adaptation. At bottom, this concept of adaptation, transported on the terrain of psychology and sociology, returns to its original meaning. It is a popular concept describing technical activity. Man adapts his tools and indirectly his organs and behavior to this material, or that situation. At the moment of its introduction into biology in the nineteenth century, the concept preserved the meaning of a relation of externality, of confrontation between an organic form and an environment opposing it, from its domain of importation. This concept has since been theoretically conceived as starting from two inverse principles, teleological and mechanist. According to one, the living being adapts itself to conform to the search for functional satisfaction; according to the other, the living being is adapted under the effect of necessities that may be mechanical, physicochemical or biological (the other living creatures in the biosphere). In the first interpretation, adaptation is the solution to a problem of an optimum forming the factual data of the environment and the living being's demands; in the second, adaptation expresses a state of equilibrium, whose lower limit defines the worst for the organism, which is the risk of death. But in both theories, the environment is considered as a physical fact, not as a biological fact, as an already constituted fact and not as a fact to be constituted. By contrast, if the organism–environment relation is considered as the effect of a really biological activity, as the search for a situation in which the living being receives, instead of submits to, influences and qualities which meet its demands, then the environments in which the living beings find themselves are carved out by them, centered on them. In this sense the organism is not thrown into an environment to which he must submit, but he structures his environment at the same time that he develops his capacities as an organism.[67]

This is particularly true of environments and modes of life peculiar to man, at the heart of technical–economic groups which, in a given geographical environment, are characterized less by the activities which are offered them than by those which they choose. Under these conditions the normal and abnormal are determined less by the encounter of two independent causal series, the organism and the environment, than by the quantity of energy at the disposal of the organic agent for delimiting and structuring this field of experiences and enterprises, called its environment. But, you will ask, where is the measure of this quantity of energy? It is to be sought nowhere other than in the history of each one of us. Each of us fixes his norms by choosing his models of exercise. The norm of a long-distance runner is not that of a sprinter. Each of us changes his norms according to his age and former norms. The norm of the former sprinter is not that of a champion. It is normal, that is, in conformity with the biological law of aging, that the progressive reduction of the margins of security involves lowering the thresholds of resistance to aggressions from the environment. The norms of an old man would have been considered deficiencies in the same man just reaching adulthood. This recognition of the individual and chronological relativity of norms is not skepticism before multiplicity but tolerance of variety. In the 1943 *Essay* we called "normativity" the biological capacity to challenge the usual norms in case of critical situations, and proposed measuring health by the gravity of the organic crises which are surmounted by the establishment of a new physiological order.[68]

In the admirable, moving pages of *The Birth of the Clinic*, Michel Foucault has shown how Bichat made "the medical gaze pivot on itself," in order to call death to account for life.[69] Not being a physiologist, we do not presume to believe that in the same way we have called disease to account for health. It is very clear that that is what we would have wanted to do so as not to hide our delight in having found after all the absolution of our former ambition in Dr. Henri Péquignot:

> In the past all the people who tried to build a science of the normal without being careful to start from the pathological considered as the immediate given have ended up in often ridiculous failures.[70]

As we are quite persuaded of the fact, analyzed above, that the knowledge of life, like the knowledge of society, assumes the priority of infraction over regularity, we would like to end these new reflections on the normal and the pathological by sketching a paradoxical pathology of the normal man, by showing that the consciousness of biological normality includes the relation to disease, the recourse to disease as the only touchstone which this consciousness recognizes and thus demands.

In what sense is the normal man's disease to be understood? Not in the sense that only the normal man can become sick, as only the ignorant can become wise. Not in the sense that slight accidents happen to disturb without nonetheless altering a state of equality and equilibrium: colds, headaches, a rash, colic, every accident without the value of a symptom, a warning without alarm.

By disease of the normal man we must understand the disturbance which arises in the course of time from the permanence of the normal state, from the incorruptible uniformity of the normal, the disease which arises from the deprivation of diseases, from an existence almost incompatible with disease. It must be admitted that the normal man knows that he is so only in a world where every man is not normal, consequently he knows he is susceptible to disease, as a good pilot knows he can run his ship aground, as an urbane man knows he can commit a blunder [*gaffe*]. The normal man feels himself capable of running his body aground, but experiences the certitude of repelling the eventuality. In the case of disease the normal man is he who lives the assurance of being able to arrest within himself what in another would run its course. In order for the normal man to believe himself so, and call himself so, he needs not the foretaste of disease but its projected shadow.

In the long run a malaise arises from not being sick in a world where there are sick men. And what if this were not because one is stronger than the disease or stronger than others, but simply because the occasion has not presented itself? And what if, in the end, when the occasion does arise, one were to show oneself as weak, as unprepared as, or perhaps more so than others? Thus there arises in the normal man an anxiety about having remained normal, a need for disease as a test of health, that is, as its proof, an unconscious search for disease, a provocation of it. Normal man's disease is the appearance of a fault in his biological confidence in himself.

Our sketch of pathology is obviously a fiction. The analyses for which it substitutes can be rapidly reconstituted with Plato's help.

> Yet that is what we say literally – we say that the physician erred and the calculator and the schoolmaster. But the truth, I take it, is, that each of these in so far as he is that which we entitle

> him never errs; so that, speaking precisely, since you are such a stickler for precision, no craftsman errs. For it is when his knowledge abandons him that he who goes wrong goes wrong – when he is not a craftsman.[71]

Let us apply what is said above of the doctor to his client. We shall say that the healthy man does not become sick insofar as he is healthy. No healthy man becomes sick, for he is sick only insofar as his health abandons him and in this he is not healthy. The so-called healthy man thus *is not* healthy. His health is an equilibrium which he redeems on inceptive ruptures. The menace of disease is one of the components of health.

Epilogue

Undoubtedly our conception of the normal is very archaic, while it is – undoubtedly because it is – as was pointed out to us in 1943, a conception of life such as can be formed when one is young. A judgment which we did not intend has delighted us and we ask to be allowed to apply it to ourselves: "The notion of this ideal which is the normal is identified with the previous euphoric state of the subject who has just fallen sick. . . . The only pathology ascertained at the time was a pathology of young subjects."[72] And undoubtedly it took the temerity of youth to believe oneself equal to the task of a study of medical philosophy on norms and the normal. The difficulty of such an undertaking makes one tremble. We are aware of this today as we complete these pages of resumption. With this confession, the reader will measure how much we, in conformity with our discussion on norms, have reduced our own with time.

Notes

Introduction

1. Fontenelle, *Préface à l'histoire de l'académie*, "Oeuvres" edition, 1790, Vol. 6, pp. 73–74. Canguilhem cites this text in his *Introduction à l'histoire des sciences*, Paris, 1970, Vol. 1, pp. 7–8.

2. On this theme, see Canguilhem's *Idéologie et rationalité dans l'histoire des sciences de la vie*, Paris, 1977, p. 21.

3. Cf. Canguilhem's *Études d'histoire et de philosophie des sciences*, Paris, 1968, p. 17.

4. Canguilhem again takes up the example dealt with by Florkin in the latter's *A History of Biochemistry*, Amsterdam, 1972–75.

5. *Idéologie et rationalité*, p. 14.

6. On the relation between epistemology and history, see in particular the "Introduction" to *Idéologie et rationalité*, pp. 11–29.

7. *Études*, p. 239.

8. Canguilhem, *La connaissance de la vie*, 2nd ed., Paris, 1965, p. 88.

9. Cf. Canguilhem's *La formation du concept de réflexe aux XVIIe et XVIIIe siècles*, Paris, 1955.

Section I

1. The references in brackets refer to the Bibliography. The first number (roman) refers to the corresponding numbered entry in the Bibliography, the

second number (in italics) to the volumes, pages or articles in the cited work.

2. On the relations between Comte and Robin, see Gentry [42] and Klein [64].

3. This text is quoted on p. 33 *in fine*. Nietzsche's lines are from *The Will to Power* (No. 47). Translated by Walter Kaufman and R. J. Hollingdale, New York, Vintage, 1968.

4. A very recently made bibliographical discovery confirms our choice. The pathological dogma we want to discuss was expounded without reticence or reservation by Charles Daremberg in 1864 in the *Journal des débats*, under the aegis of Broussais, Comte, Littré, Charles Robin and Claude Bernard [29].

5. For Comte's biological and medical reading between 1817 and 1824, when "he was preparing to become not a biologist but a philosopher of biology," see H. Gouhier [47, *237*].

6. A good exposition of all Broussais's ideas can be found in [14; 29; 13 *bis*, *III*; 83].

7. Italics ours.

8. Italics ours.

9. [The following English versions of Brown's *Elementa medicinae* are taken from the 1806 Philadelphia edition; the page references are from that edition – *Trans.*]

10.* Cf. my recent study "John Brown, La théorie de l'incitabilité de l'organisme et son importance historique." In the *Proceedings of the XIII International Congress of the History of Science*, Section IX, *The History of Biology* (including the fundamentals of medicine), Moscow, Editions Nauka, 1974, pp. 141–146.

11. Bernard willed his unpublished papers to d'Arsonval. Cf. *Cl. Bernard, Pensées, notes détachées*, preface by d'Arsonval (J.-B. Baillière, 1937). These papers have been inventoried by Delhoume but so far he has only published fragments.

* Today a *Catalogue des manuscripts de Cl. Bernard* is available, edited by M.-D. Grmek, Paris, Masson, 1967.

12. Pasteur in the article on *Cl. Bernard, ses travaux, son enseignement, sa méthode* [93].

13. See Pierre Mauriac's *Claude Bernard* [81] and Pierre Lamy's *Claude Bernard et le matérialisme* [69].

14. For example, this is the case of Henri Roger in his *Introduction à la médecine*, Paris, G. Carré et C. Naud, 1899. The same is true of Henri Claude and Jean Camus in their *Pathologie générale*, Paris, J.-B. Baillière et fils, 1909.

15. Physiology course on *La constance du milieu intérieur* [The Constancy of the Interior Environment], Faculty of Medicine, Toulouse, 1938–39.

16. Course in pharmacology, Strasbourg Faculty of Medicine, 1941–42.

17. This expression, "inconspicuous infection," seems to me to be incorrect. Infection is invisible only from the clinical point of view and on the macroscopic level. But from the biological point of view and on the humoral level, infection is apparent since it can be expressed in terms of the presence of antibodies in the serum. Nevertheless, infection is only a biological fact, a modification of the humors. An inapparent infection is not an inapparent disease.

18. Bernard says that he never succeeded in detecting sugar in the tears of a diabetic, but today it is an established fact; cf. Fromaget and Chaix, "Glucides," *Physiologie*, fasc. 3, year 2, p. 40, Paris, Hermann, 1939.

19.* After the original publication of this study (1943) an examination of Bernard's ideas was undertaken by Dr. M.-D. Grmek, "La conception de la santé et de la maladie chez Claude Bernard" [cf. Bibliography to Section II].

20. *Nature and Life*, Cambridge [Eng.], 1934, p. 5. Quoted by Koyré in a report in *Recherches philosophiques* IV (1934–35), 398.

21. Besides, Hegel understood this perfectly well, cf. *Wissenschaft der Logik* (Chapter I, 3).

22.* These questions were studied later by de Laet and Lobet, *Étude de la valeur des gestes professionels* [Study of the Value of Professional Gestures], Brussels, 1949, and by A. Geerts, *L'indemnisation des lésions corporelles à travers les âges* [Compensation for Bodily Injuries Through the Ages], Paris, 1962.

23. In his *Dictionnaire étymologique grec et latin* (1942) A. Juret proposes this same etymology for the word "anomaly."

24. This is the actual expression used by Flourens.

25. Report of the Danish Expedition on the North East Coast of Greenland 1906–1908. *Meddelelser om Grønland*, p. 44, Copenhagen, 1917. Quoted after R. Isenschmidt, "Physiologie der Wärmeregulation," in *Handbuch der norm. u. path. Physiologie* 17, p. 3, Berlin, Springer, 1926.

26.* We no longer take the liberty of asking ourselves this today.

27. *Wertphilosophie und Ethik*, p. 29, 1939, Vienna-Leipzig, Braumüller.

28. This number of 40 pulsations seems less extraordinary than Sigerist's example gives us to understand, when one is acquainted with the influence of athletic training on cardiac rhythm. The pulse decreases in frequency as training advances. This decrease is more pronounced in a thirty-year-old subject than in a twenty year old. It also depends on the kind of sport engaged in. In an oarsman a pulse of 40 indicates very good form. If the pulse falls below 40, one speaks of overconditioning.

29. Merleau-Ponty's work, *Structure du comportement* (Alcan, 1942), has just done a great deal to spread Goldstein's ideas.

* A French translation of [Goldstein's] *Aufbau des organismus* by E. Burckardt and J. Kuntz appeared in 1951 (Gallimard) under the title *La structure de l'organisme*.

30. [There is a play on words in the French: "... médecin qui dit: Ménagez-vous! 'Me ménager c'est bien facile à dire, mais j'ai mon ménage [household]' "; p. 130 – *Trans.*]

31.* On this point see W. B. Cannon, *The Wisdom of the Body*, Chapter XIV: "The Margin of Safety in Bodily Structure and Functions," New York, 1932.

32. It may be objected, perhaps, that we have a tendency to confuse health and youth. However, we are mindful of the fact that old age is a normal stage of life. But at the same age one old man will be healthy who demonstrates a capacity to adapt or repair organic damages which another may not demonstrate, for example, a good, solid knitting of the neck of a fractured femur. The beautiful old man is not merely poetic fiction.

33. Singer in his otherwise noteworthy section devoted to Harvey insists rather on the traditional character of his biological conceptions, so that Harvey

would have been an innovator because of his methodological correctness and despite his doctrinal assumptions [108].

34.* Cf. M.-D. Grmek's study "Opinion de Claude Bernard sur Virchow et la pathologie cellulaire," in *Castalia* (Milan), January–June 1965.

35. Circumstances have not allowed us to refer directly to Ricker's works.

36.* In the Soviet Union it is A.-D. Speransky, *Fondements de la théorie de la médecine*, 1934 [English translation, New York, International Publishers, 1936; German translation, 1950]. Cf. Jean Starobinski's study, "Une théorie de l'origine nerveuse des maladies," *Critique*, No. 47, April 1951.

37. It is understood that we are not dealing here with mental illnesses where the patients' ignorance of their state often constitutes an essential aspect of the disease.

SECTION II

1. Francis Courtès, *maître-assistant* at the Montpellier Faculty of Letters and Social Sciences.

2. Cf. *supra*, pp. 78–86.

3. *Principes de médecine expérimentale*, p. 170.

4. Cf. my article "La pensée de René Leriche," in *Revue philosophique* 146 (1956), 313–317.

5. Ch. Kayser, "Le maintien de l'équilibre pondéral," *Acta neurovegetativa* 24, 1–4, Vienna, 1963.

6. *La terre et les rêveries du repos* (Paris, 1948), pp. 41–42.

7. It would be possible and fruitful – but this is not the place – to set up semantic families of concepts representing the kinship of normal and abnormal, for example, the set *tur*bid, *tor*tured, *tor*tuous, etc., and the set *oblique, deviated, athwart*, etc.

8. *Tristes tropiques*, 38, "A Little Glass of Rum."

9. "Aux origines de la pensée sociologique," *Les temps modernes* (December 1962).

10. Descartes had already said: "Although health is the greatest of all our bodily possessions, it is nevertheless the one about which we reflect the least

and which we savor least. The knowledge of the truth is like the health of the soul: as long as we have it we don't think any more about it." (Letter to Chanut, 31 March 1649.)

11. *The Sophist*, 239b, in *The Sophist and the Statesman*, translation and introduction by A. E. Taylor. Edited by R. Klibansky and E. Anscombe (London, Nelson, 1961). [Orthology: the art of using words correctly (*Webster's New International Dictionary*, 2nd ed. 1958).]

12. Cf. Pierre Guiraud, *La grammaire*, Presses Universitaires de France, 1958, p. 109.

13. *Remarques sur la langue française* (1647), preface.

14. Establishment of conscription and the medical examination of conscripts; establishment of national studfarms and remount depots.

15. *Op. cit.*, p. 109.

16. Cf. Jacques Maily, *La normalisation* (Paris, Dunod, 1946), pp. 157 sq. Our brief account of normalization owes much to this work which is useful for its clarity of analysis and historical information as well as for its references to a study of Dr. Hellmich, *Vom Wesen der Normung* (1927).

17. *Fables*, VI, 4, "Jupiter et le Métayer" [Jupiter and the Sharecropper].

18. Julien Freund, *L'essense du politique* (Paris, Sirey, 1965), p. 332.

19. *Op. cit.*, p. 293

20. Cf. Bergson, *The Two Sources of Morality and Religion*: "Whether human or animal, a society is an organization; it implies a coordination and generally also a subordination of elements; it therefore exhibits, whether merely embodied in life or, in addition, specifically formulated, a collection of rules and laws." (Trans. by R. Ashley Audra and Cloudesley Brereton, Garden City, New York, Doubleday, 1954, p. 27.)

21. *Cour phil. pos.*, 48ᵉ Leçon (ed. by Schleicher, v. IV, p. 170).

22. *Ibid.*, p. 177.

23. *Ibid.*, p. 176.

24. *Ibid.*, p. 183.

25. *Syst. de pol. pos.*, II, p. 304.

26. *Ibid.*, p. 335.

27. Pp. 148–155 of that work. Is it uninteresting to recall that at the end of the nineteenth century the French Army's intelligence department, disagreeably implicated in the Dreyfus affair, bore the name of department of statistics.

28. *Le geste et la parole*: Vol. I: *Technique et langage* (Paris, 1964), p. 114.

29. *Le geste et la parole*: Vol. II: *La mémoire et les rythmes* (Paris, 1965), p. 34.

30. *Ibid.*, p. 63.

31. *Tristes tropiques*, 38, p. 387.

32. London, Cassell, 5th ed., 1910.

33. *Ibid.*, p. 5.

34. *Ibid.*, p. 6.

35. We have commented at greater length on these reflections of Chesterton in our lecture: "Le problème des régulations dans la société" (*Cahiers de l'Alliance Israélite Universelle*, No. 92, Sept.–Oct. 1955).

36. *Op. cit.*, p. 5.

37. *Out of the Night: A Biologist's View of the Future* (New York, Vanguard, 1935).

38. *Ibid.*, p. 122.

39. Cf. Julian Huxley, *Soviet Genetics and World Science: Lysenko and the Meaning of Heredity* (London, Chatto and Windus, 1949).

40. Cannon borrowed the title *The Wisdom of the Body* from the illustrious English physiologist, Starling.

41. *Ibid.*, p. 24.

42. Cf. *supra* pp. 142–144.

43. Duyckaerts, *La notion de normal en psychologie clinique* (Paris, Vrin, 1954), p. 157.

44. Published by Gallimard, 1st edition 1949; 2nd edition, 1958. The thesis of evolution by dichotomy (split of an animal group into an innovative and a conservative branch) is taken up again by Vandel in his article on "L'évolutionnisme de Teilhard de Chardin," *Études philosophiques*, 1965, No. 4, p. 459.

45. "In Darwin's terminology the vacant places in a given locality are less free spaces than systems of life (habitat, mode of feeding, attack, protection) which are theoretically possible there and not yet practiced." "Du développement

à l'évolution au XIXe siècle," by Canguilhem, Lapassade, Piquemal, Ulmann, in *Thales* 11 (1960), 32.

46. We are borrowing the core of our information on genetic homeostasis from Ernest Bösiger's excellent study, "Tendances actuelles de la génétique des populations," published in the Report of the XXVI *Semaine de Synthèse, La biologie, acquisitions récentes* (Paris, Aubier, 1965).

47. We could even say with A. Lwoff: "The living organism has no problems; in nature there are no problems; there are only solutions," "Le concept d'information dans la biologie moléculaire," in *Le concept d'information dans la science contemporaine* (Paris, Les Editions de Minuit, 1965), p. 198.

48. *Quarterly Bull. Northwestern University Medical School* 18 (Chicago, 1944), 22–32, Spring Quarter. This article was drawn to our attention and obtained for us by Profs. Charles Kayser and Bernard Metz.

49. *The Lancet* I, 1 (1947); the article is reproduced in *Concepts of Medicine*, edited by Brandon Lush (Pergamon Press, 1961).

50. Cf. *supra*, pp. 165–172.

51. *Tensions Affecting International Understanding: A Survey of Research* (New York, Social Science Research Council, 1950), pp. 46–48. This work was brought to our attention by Robert Pagès.

52. Cf. Selye: "D'une révolution en pathologie," *La nouvelle revue française* (1 March 1954), p. 409. Selye's principal work is *Stress*, Montreal, 1950. Previously, "Le syndrome général d'adaptation et les maladies de l'adaptation," *Annales d'endocrinologie* (1946), Nos. 5 and 6.

53. *Principes de physiopathologie et de thérapeutique générales* [Principles of General Physiological Pathology and Therapeutics], 3rd edition (Paris, Masson, 1963), p. 232.

54. Cf. Charles Kayser: "The study of hyperventilation at high altitudes and in the course of work has led to a serious revision of our conceptions of the importance of reflex mechanisms in regulating breathing. The importance of the heart's debit in the circulatory mechanism appeared in all its clarity only when sportsmen and sedentary people had been studied exerting themselves.

Sport and work pose a set of purely physiological problems which we must try to clarify." (*Physiologie du travail et du sport*, Paris, Hermann, 1947, p. 233.)

55. *La raison et les remèdes* (Paris, Presses Universitaires de France, 1964), p. 310.

56. Cf. on this subject Philippe Decourt, "Phénomènes de Reilly et syndrome général d'adaptation de Selye," *Études et Documents*, I (Tangier, Hesperis, 1951).

57. "Les troubles fonctionnels en pathologie" (Opening lecture in the course on medical pathology), in *La Presse médicale*, 103 (23 December 1936). This text was reported to us by François Dagognet.

58. *Op cit.*, p. 202.

59. Cf. *supra*, pp. 78–79.

60. *Inborn Errors of Metabolism* (London, H. Frowde, 1909).

61. Cf. Tubiana, "Le goitre, conception moderne," *Revue française d'études cliniques et biologiques* (May 1962), pp. 469–476.

62. For a classification of genetic diseases, cf. P. Bugard, *L'état de maladie*, Part IV (Paris, Masson, 1964).

63. On this point cf. R. Ruyer, *La cybernétique et l'origine de l'information* (Paris, 1954); and G. Simondon, *L'individu et sa genèse physico-biologique* (Paris, 1964), pp. 22–24.

64. Cf. S. Bonnefoy, *L'intolérance héréditaire au fructose* (Medical thesis, Lyon, 1961).

65. *Traité des maladies qu'il est dangereux de guérir*, by Dominique Raymond (1757). New enlarged edition with notes by M. Giraudy (Paris, 1808).

66. *L'inadaptation, phénomène social (Recherches et débats du C.C.I.F.*, Fayard, 1964), p. 39. As can be seen from Dr. Péquignot's contribution to the above-mentioned discussion on inadaptation, he does not identify abnormal and unfitness, and our critical reservations in the following lines do not concern him.

67. Cf. our study "Le vivant et son milieu" in *La connaissance de la vie* (Paris, 1965).

68. Cf. *supra*, pp. 198–199.

69. *Op cit.*, p. 146.

70. *Initiation à la médecine* (Paris, Masson, 1961), p. 26.

71. *The Republic*, with an English translation by Paul Shorey (Cambridge, Mass., Harvard University Press, 1963), Book I, 340d (pp. 55 and 57).

72. H. Péquignot, *Initiation à la médecine*, p. 20.

Glossary of Medical Terms

This Glossary was prepared by Carolyn Fawcett and Robert Cohen. It is based in part on *Stedman's Medical Dictionary* (22nd ed., Baltimore, Williams and Wilkins, 1975).

achlorhydria: the absence of hydrochloric acid from the gastric juice.

adenoma: a benign tumor of epithelial tissue forming a gland or glandlike structure.

alkaptonuria: an acid secretion in the urine due to a specific, congenital enzymatic lack.

alveolar pyorrhea: periodontitis.

anaphylaxia: lessened resistance and extreme sensitivity of tissues to the reintroduction of foreign protein or other material.

ankylosis: stiffening or fixation of a joint.

aphasia: loss or impairment of the power to use or understand language in any of its forms, resulting from a brain lesion, or sometimes from functional or emotional disturbance.

apnea: absence of respiration.

apophysis: an outgrowth or projection, especially from a bone.

areflexia: a condition in which the reflexes are absent.

arteritis: inflammation involving the arteries.

asthenia: weakness, debility.

autophagia: feeding upon oneself; maintenance of the nutrition of the whole body by metabolic consumption of some of the body tissues.

axis cylinder: see *neurite*.

Basedow's disease: (also *Graves's disease*): toxic goiter (enlargement of the thyroid), sometimes but not always accompanied by protrusion of the eyeballs.

calcemia: excess of calcium in the blood.

cardiac liver (*cardiac cirrhosis*): a liver disease resulting from prolonged congestive heart failure.

cenesthesia: the general sense of bodily existence related to the functioning of the internal organs.

convoluted tubule: convoluted portion of the functional unit of the kidney.

Cooley's anemia (*thalassemia major*): severe anemia due to an inherited disorder of hemoglobin metabolism.

coxalgia (*coxodynia*): pain in the hip joint.

cupping glasses: glasses used in cupping (a method once used to produce slow bleeding by applying a glass cup to an opening in the skin to create a partial vacuum).

cyanotic: relating to or marked by cyanosis, a dark bluish or purplish coloration of the skin and mucous membrane due to deficient oxygenation of the blood.

cyclopia: a congenital defect in which the two eye orbits merge to form a single cavity containing one eye.

cystinuria: excessive urinary excretion of cystine, an amino acid occurring in protein, notably keratin and insulin.

diastase: an enzymatic mixture which converts starch into dextrin and maltose.

diathesis: a permanent (hereditary or acquired) condition of the body which renders it liable to certain special diseases or metabolic or structural anomaly.

diuresis: excretion of urine, usually denoting large volume.

ectromelia: a congenital lack of one or more of the limbs.

edema: an accumulation of an excessive amount of fluid in cells, tissues or serous cavities.

endocarditis: inflammation of the lining membrane of the heart.

epiphysis: a part of a long bone which ossifies separately from that of the shaft and which subsequently becomes fused to the main part of the bone, in some cases in man as late as the twentieth year.

erysipelas: a local febrile disease accompanied by diffused inflammation of the skin, producing a deep red color.

exostosis: a bony tumor springing from the surface of a bone.

gastrectomy: removal of part or all of the stomach.

glycemia: presence of glucose in the blood.

glycogenesis: process of the formation of glycogen from glucose.

glycosuria: urinary excretion of carbohydrates.

hemeralopia: day blindness, or more generally, the inability to see as distinctly in a bright light as in a dim one.

hemianopsia: loss of vision for one half of the visual field of one or both eyes.

hepatic aldolases: a class of enzymes (cleaving carbon bonds of aldohydes) found in the liver.

Hodgkin's disease: chronic enlargement of the lymph nodes due to a malignant tumor of reticulum cells.

hyperchlorhydria: the presence of an abnormal amount of hydrochloric acid in the stomach.

hyperglycemia: abnormally high concentration of glucose in the blood.

hypermyotonia: extreme muscular tonus.

hyperthermia: unusually high fever.

hypertonia: extreme tension of the muscles or arteries.

hypophysectomy: removal of the pituitary gland.

hypospadias: a developmental anomaly in the wall of the urethra which then opens, in males, at a distance under the surface of the penis, and in females, directly into the vagina.

hypothermia: body temperature below 98.6 F°.

integumentary: relating to the enveloping membrane of the body (integument).

isles of Langerhans: cellular masses in the interstitial tissue of the pancreas, composed of different cell types and comprising the endocrine portion of the pancreas; they are the source of insulin and glucagon.

leukocyte: white blood cell.

Löffler bacillus: *corynebacterium diphtheriae* which, along with its highly potent toxin, causes diphtheria.

miasma: noxious exhalations formerly regarded as the cause of malaria and of various epidemic diseases.

mongolian trisomy: trisomy refers to an individual or cell with an extra chromosome; in man a trisomic cell has forty-seven chromosomes. Trisomy characterizes the cells of almost all mongoloids.

necrosis: the pathologic death of one or more cells, or of a portion of tissue or of an organ.

nephritis: inflammation of the kidneys.

neurite: axon.

neuroglia: non-nervous cellular elements of nervous tissue, with supporting and possibly metabolic functions.

neuroglioma: a tumor developed from neuroglial cells.

neurolemma (or *sheath of Schwann*): a tubular cell that enfolds some nerve fibers.

neuroma: the old, general term for any neoplasm derived from cells of the nervous system.

pancreatectomy: removal of the pancreas.

paraplegia: paralysis of both legs and generally the lower trunk as well.

parenchyma: the distinguishing or specific cells of an organ as distinct from the connective tissue framework.

Parkinson's disease (*paralysis agitans*): a neurological disorder caused by degenerative disease of the basal ganglia, associated with rigidity, tremor, poverty of movement, odd posture, peculiar acceleration of gait.

pentosuria: excretion of one or more pentoses (simple sugars containing five carbon atoms) in the urine.

peripneumonia: old name for pneumonia.

phenylketonuria: an inherited metabolic deficiency resulting in brain damage with severe mental retardation and neurological abnormalities.

phenylpyruvic imbecility: see *phenylketonuria*.

phenylpyruvic oligophrenia: see *phenylketonuria*.

phlegmasia: inflammation, especially when acute and severe.

phthisis: a wasting or atrophy; in particular (as used in this book) an obsolete term for consumption or tuberculosis of the lungs.

pithiatism: morbid condition curable by suggestion.

polydipsia: frequent drinking because of extreme thirst.

polyphagia: excessive eating.

polyuria: excessive excretion of urine.

Pott's disease (*tuberculous spondylitis*): tuberculous infection of the spine.

pronation of the forearm: rotation of the forearm so that the palm faces backward when the arm is at the side of the body.

pylorus: a muscular or myovascular device to open and close an orifice: in Canguilhem's example, the opening from the stomach into the intestine.

reticulo-endothelial system: system of large macrophages or mononucleated cells found in the linings of the sinuses and in the spleen, liver, lymph nodes, bone marrow, connective system, etc. which ingest dead tissue and degenerated cells, and form part of the body's immunity system.

sacralization: an anomaly of the fifth lumbar vertebra involving fusion with the upper part of the sacrum (next to the lowest part of the spinal column).

scleroderma: a disease characterized by swelling and thickening of the skin (hidebound disease).

sclerosis: a morbid hardening of any tissue or structure of chronic inflammatory origin.

splanchnic: one of the nerves supplying the viscera.

spondylosis: breaking down or dissolution of the body of a vertebra.

stenosal: related to the narrowing of any canal.

sthenia: a condition of activity and apparent force.

supination of the forearm: rotation of the forearm so that the palm faces upward when the arm is at the side of the body.

syringomyelia: a chronic disease involving cavities in the spinal chord, characterized by muscular atrophy, especially of the upper extremities, loss of the senses of pain and temperature with tactile sensibility retained.

tabes: progressive wasting or emaciation.

terrain (adapted from Robert): the state of an organism with regard to its resistance to pathogenic agents or its predisposition to different diseases.

tetanus: a disease marked by painful tonic muscular contractions.

tonus: a state of normal tension of the tissues by virtue of which the parts are kept in shape and alert and ready to function.

tryptophane: a component of proteins.

tyrosine: a specific amino acid present in most proteins.

uremia: excess of urea and other nitrogenous waste in the blood.

vasodilatation: dilation of the blood vessels.

Bibliographies

In the text the references in brackets refer to this Bibliography. The first number (roman) refers to the corresponding number entry in this Bibliography; the second number (in italics) to the volumes, pages or articles in the cited work.

Section I

1. Abelous, J.-E., "Introduction à l'étude des sécrétions internes," *Traité de physiologie normale et pathologique,* 2nd ed., Vol. IV. Paris, Masson, 1939.
2. Ambard, L., "La biologie," *Histoire du monde,* edited by E. Cavaignac, Vol. XIII, Part V. Paris, de Boccard, 1930.
3. Bégin, L.-J., *Principes généraux de physiologie pathologique coordonnés d'après la doctrine de M. Broussais.* Paris, Méquignon-Marvis, 1821.
4. Bernard, Cl., *Leçons de physiologie expérimentale appliquée à la médicine,* 2 vols. Paris, J.-B. Baillière, 1855–1856.
5. ——, *Leçons sur les propriétés physiologiques et les altérations pathologiques des liquides de l'organisme,* 2 vols. Paris, J.-B. Baillière, 1859.
6. ——, *Introduction à l'étude de la médecine expérimentale.* Paris, J.-B. Baillière, 1865. (*Introduction to the Study of Experimental Medicine.* Translated by Henry Copley Greene. New York, Macmillan, 1927; New York, Collier, 1961.)
7. ——, *Rapport sur les progrès et la marche de la physiologie générale en France.* Paris, Imprimerie impériale, 1867.
8. ——, *Leçons sur la chaleur animale.* Paris, J.-B. Baillière, 1876.
9. ——, *Leçons sur le diabète et la glycogenèse animale.* Paris, J.-B. Baillière, 1877.
10. ——, *Leçons sur les phénomènes de la vie communs aux animaux et aux végétaux.* 2 vols. Paris, J.-B. Baillière, 1878–79. (*Lectures on the Phenomena of Life Common to Animals and Plants.* Translated by Hebbel E. Hoff, Roger Guillemin and Lucienne Guillemin. Springfield, Ill., Thomas, 1974.)

11. ——, *Philosophie; manuscrit inédit*. Paris, Boivin, 1938.

12. Bichat, X., *Recherches sur la vie et la mort*. Paris, Béchet, 1800; 4th ed. with notes by Magendie, 1822. (*Physiological Researches on Life and Death*. Translated by F. Gold, with notes by Magendie translated by George Hayward. Boston, Richardson and Lord, 1827.)

13. ——, *Anatomie générale appliquée à la physiologie et à la médecine*. Paris, Brosson and Chaudé, 1801; new ed. by Béclard, 1821. (*General Anatomy, Applied to Physiology and Medicine*. Translated by George Hayward, 2 vols. Boston, Richardson and Lord, 1822.)

13bis. de Blainville, C., *Histoire des sciences de l'organisation et de leurs progrès comme base de la philosophie*. Paris, Périsse, 1845. (In Volume II, see Haller; in Volume III, see Pinel, Bichat, Broussais.)

14. Boinet, E., *Les doctrines médicales. Leur évolution*. Paris, Flammarion, n.d.

15. Bordet, J., "La résistance aux maladies." *Encyclopédie française*, Vol. 6, 1936.

16. Bounoure, L., *L'origine des çellules reproductrices et le problème de la lignée germinale*. Paris, Gauthier-Villars, 1939.

17. Brosse, Th., "L'énergie consciente, facteur de régulation psycho-physiologique," *Evolution psychiatrique* 1 (1938). (See also Laubry and Brosse [70].)

18. Broussais, F.-J.-V., *Traité de physiologie appliquée à la pathologie*, 2 vols. Paris, Mlle Delauney, 1822–23. (*A Treatise on Physiology Applied to Pathology*. Translated by John Bell and R. LaRoche. Philadelphia, H.C. Carey and I. Lea, 1826.)

19. ——, *Catéchisme de la médecine physiologique*. Paris, Mlle Delauney, 1824.

20. ——, *De l'irritation et de la folie*. Paris, Mlle Delauney, 1828. (*On Irritation and Insanity*. . . . Translated by Thomas Cooper. Columbia, S.C., S. J. M'Morris, 1831.)

21. Brown, John, *Elementa medicinae*. Edinburgh, C. Elliot, 1780–84, 2 vols. (*The Elements of Medicine*. London, J. Johnson, 1788, 2 vols. in 1; *The Elements of Medicine*. Philadelphia, printed by A. Bartram for Thomas Dobson, 1806; *Eléments de médecine*. . . *traduits de l'original latin*. . . *avec la table de Lynch par Fouquier*. Paris, Demonville et Gabon, 1805.)

22. Cassirer, E., "Pathologie de la conscience symbolique," *Journal de psychologie* 26 (1929), 289–336 and 523–566.
23. Castiglioni, A., *Storia della medicina.* Milan, Società editrice "Unitas," 1927; new, enlarged and updated ed. Milan, Mondadori, 1948, 2 vols. (*A History of Medicine.* Translated by E.B. Krumbhaar. New York, Knopf, 1941; 2nd rev. and enl. edition, New York, Knopf, 1958.)
24. Caullery, M., *Le problème de l'evolution.* Paris, Payot, 1931.
25. Chabanier, H. and Lobo-Onell, C., *Précis du diabète.* Paris, Masson, 1931.
26. Comte, A., "Examen du traité de Broussais sur l'irritation," 1828. Appendix to the *Système de politique positive* (cf. [28]), Vol. IV, p. 216.
27. —, "Considérations philosophiques sur l'ensemble de la science biologique," 1838. 40th lecture of the *Cours de philosophie positive.* Paris, ed. by Schleicher, 1908. Vol. III.
28. —, *Système de politique positive,* 4 vols. Paris, Crès, 1851–54; 4th ed., 1912. (*System of Positive Polity.* London, Longman, Green and Co., 1875–77, 4 vols.)
29. Daremberg, Ch., *La médecine, histoire et doctrines.* 2nd ed., Paris, J.-B. Baillière, 1865. ("De la maladie," p. 305.)
30. —, *Histoire des sciences médicales,* 2 vols. Paris, J.-B. Baillière, 1870.
31. Déjerine, J., *Sémiologie des affections du système nerveux.* Paris, Masson, 1914.
32. Delbet, P., "Sciences médicales." In *De la méthode dans les sciences, I,* by Bouasse, Delbet, etc. Paris, Alcan, 1909.
33. Delmas-Marsalet, P., *L'électrochoc thérapeutique et la dissolution-reconstruction.* Paris, J.-B. Baillière, 1943.
34. Donald C. King, "Influence de la physiologie sur la littérature française de 1670 à 1870." Thesis, Paris, 1929.
35. Dubois, R., *Physiologie générale et comparée.* Paris, Carré and Naud, 1898.
36. Duclaux, J., *L'analyse physico-chemique des fonctions vitales.* Paris, Hermann, 1934.
37. Dugas, L., *Le philosophe Théodule Ribot.* Paris, Payot, 1924.
38. Ey, H. and Rouart, J., "Essai d'application des principes de Jackson à une conception dynamique de la neuro-psychiatrie," *Encéphale* 31 (1936), 313–356.

39. Flourens, P., *De la longévité humaine et de la quantité de vie sur le globe*. Paris, Garnier, 1854; 2nd ed., 1855. (*On Human Longevity, and the Amount of Life Upon the Globe*. Translated from the 2nd French edition by Charles Martel. London, H. Baillière, 1855.)

40. Frédéricq, H., *Traité élémentaire de physiologie humaine*. Paris, Masson, 1942.

41. Gallais, F., "Alcaptonurie." In *Maladies de la nutrition, Encyclopédie médico-chirurgicale*. 1st ed. 1936.

42. Gentry, V., "Un grand biologiste: Charles Robin, sa vie, ses amitiés philosophiques et littéraires." Thesis in medicine, Lyon, 1931.

43. Geoffroy Saint-Hilaire, I., *Histoire générale et particulière des anomalies de l'organisation chez l'homme et les animaux*. 3 vols. and one atlas. Paris, J.-B. Baillière, 1832.

44. Gley, E., "Influence du positivisme sur le développement des sciences biologiques en France," *Annales internationales d'histoire*. Paris, Colin, 1901.

45. Goldstein, K., "L'analyse de l'aphasie et l'étude de l'essence du langage," *Journal de Psychologie* 30 (1933), 430–496.

46. ——, *Der Aufbau des Organismus*. The Hague, Nijhoff, 1934. (*The Organism, a Holistic Approach to Biology Derived from Pathological Data in Man*. New York, American Book Co., 1939; Boston, Beacon Press, 1963.)

47. Gouhier, H., *La jeunesse d'A. Comte et la formation du positivisme*: III, *A. Comte et Saint-Simon*. Paris, Vrin, 1941.

48. Guardia, J.-M., *Histoire de la médecine d'Hippocrate à Broussais et ses successeurs*. Paris, Doin, 1884.

49. Gurwitsch, A., "Le fonctionnement de l'organisme d'après K. Goldstein," *Journal de Psychologie* 36 (1939), 107–138.

50. ——, "La science biologique d'après K. Goldstein," *Revue philosophique* 129 (1940), 244–265.

51. Guyénot, E., *La variation et l'évolution*, 2 vols. Paris, Doin, 1930.

52. ——, "La vie comme invention." In *L'Invention, 9ᵉ semaine internationale de synthèse*. Paris, Alcan, 1938.

53. Halbwachs, M., "La théorie de l'homme moyen: Essai sur Quetelet et la statistique morale." Thesis, Paris, 1912.

53[bis.] Hallion, L. and Gayet, R., "La régulation neurohormonale de la glycémie." In *Les Régulations hormonales en biologie, clinique et thérapeutique.* Paris, J.-B. Baillière, 1937.

54. Hédon, L. and Loubatières, A., "Le diabète expérimental de Young et le rôle de l'hypophyse dans la pathogène du diabète sucré," *Biologie médicale* (March–April 1942).

55. Herxheimer, G., *Krankheitslehre der Gegenwart. Strömungen und Forschungen in der Pathologie seit 1914.* Dresden-Leipzig, Steinkopff, 1927.

56. Hovasse, R., "Transformisme et fixisme: Comment concevoir l'évolution?" *Revue médicale de France* (January–February 1943).

57. Jaccoud, S., *Leçons de clinique médicale faites à l'Hôpital de la Charité.* Paris, Delahaye, 1867.

58. ——, *Traité de pathologie interne.* Vol. III, 7th ed. Paris, Delahaye, 1883.

59. Jaspers, K., *Allgemeine Psychopathologie.* Berlin, Springer, 1913; 3rd enl. and rev. ed. Berlin, Springer, 1923. (*General Psychopathology.* Translated from the 7th German ed. by J. Hoenig and Marian W. Hamilton. Chicago, University of Chicago Press, 1963.)

60. Kayser, Ch., (with A. Ginglinger), "Établissement de la thermorégulation chez les homéothermes au cours du développement," *Annales de physiologie* 5 (1929), No. 4.

61. ——, (with E. Burckardt and L. Dontcheff), "Le rythme nycthéméral chez le pigeon," *Annales de physiologie* 9 (1933), No. 2.

62. ——, (with L. Dontcheff), "Le rythme saisonnier du métabolisme de base chez le pigeon en fonction de la température moyenne du milieu," *Annales de physiologie* 10 (1934), No. 2.

63. ——, (with L. Dontcheff and P. Reiss), "Le rythme nycthéméral de la production de chaleur chez le pigeon et ses rapports avec l'excitabilité des centres thermorégulateurs," *Annales de physiologie* 11 (1935), No. 5.

63[bis.] ——, "Les réflexes." In *Conférences de physiologie médicale sur des sujets d'actualité.* Paris, Masson, 1933.

64. Klein, M., *Histoire des origines de la théorie cellulaire.* Paris, Hermann, 1936. (See also Weiss and Klein [119].)

65. Labbé, M., "Etiologie des maladies de la nutrition." In *Maladies de la nutrition, Encyclopédie médico-chirurgicale,* 1936. 1st ed.
66. Lagache, D., "La méthode pathologique," *Encyclopédie française* 8, 1938.
67. Lalande, A., *Vocabulaire technique et critique de la philosophie.* 2 vols. and 1 suppl. 4th ed. Paris, Alcan, 1938.
68. Lamy, P., "L'Introduction à l'étude de la médecine expérimentale. Claude Bernard, le Naturalisme et le Positivisme." Thesis, Paris, 1928.
69. ——, *Claude Bernard et le matérialisme.* Paris, Alcan, 1939.
70. Laubry, Ch. and Brosse, Th., "Documents recueillis aux Indes sur les 'Yoguis' par l'enregistrement simultané du pouls, de la respiration et de l'électrocardiogramme." *La Presse médicale,* 14 Oct. 1936.
71. Laugier, H., "L'homme normal," *Encyclopédie française* 4, 1937.
72. Leriche, R., "Recherches et réflexions critiques sur la douleur," *La Presse médicale,* 3 Jan. 1931.
73. ——, "Introduction générale"; "De la santé à la maladie"; "La douleur dans les maladies"; "Où va la médecine?" *Encyclopédie française* 6, 1936.
74. ——, *La chirurgie de la douleur.* Paris, Masson, 1937; 2nd ed., 1940. (*The Surgery of Pain.* Translated and edited by Archibald Young. Baltimore, Williams and Wilkins, 1939.)
75. ——, "Neurochirurgie de la douleur," *Revue neurologique* 68 (1937), 317–342.
76. ——, *Physiologie et pathologie du tissu osseux.* Paris, Masson, 1939.
76bis. Lefrou, G., *Le Noir d'Afrique.* Paris, Payot, 1943.
77. L'Héritier, Ph. and Teissier, G., "Discussion du Rapport de J.-B. S. Haldane: 'L'analyse génétique des populations naturelles.' " In *Congrès du Palais de la Découverte,* 1937: *VIII, Biologie.* Paris, Hermann, 1938.
78. Littré, E., *Médecine et médecins.* Paris, Didier, 1872. 2nd ed.
79. ——, and Robin, Ch., *Dictionnaire de médecine, chirurgie, pharmacie, de l'art vétérinaire et des sciences qui s'y rapportent.* 13th ed. completely revised. Paris, J.-B. Baillière, 1873.
80. Marquezy, R.-A. and Ladet, M., "Le syndrome malin au cours des toxi-infections. Le rôle du système neuro-végétatif." In *Xe Congrés des Pédiatres de Langue française.* Paris, Masson, 1938.

81. Mauriac, P., *Claude Bernard.* Paris, Grasset, 1940.

82. Mayer, A., "L'organisme normal et la mesure du fonctionnement," *Encyclopédie française* 4. Paris, 1937.

83. Mignet, M., "Broussais." In *Notices et portraits historiques et littéraires.* Vol. I, 3rd ed. Paris, Charpentier, 1854.

84. Minkowski, E., "A la recherche de la norme en psychopathologie," *Évolution psychiatrique* (1938), No. 1.

85. Morgagni, A., *De sedibus et causis morborum . . .*, 2 vols. Venice, Ex typographia remondiniana 1761. (*The Seats and Causes of Diseases* .... Translated by Benjamin Alexander, 3 vols. London, printed for A. Millar and T. Cadell, his successor, etç., 1769; reprinted New York, Hafer, 1960.) See Vol. I for the Dedicatory epistle of 31 August 1760.

86. Mourgue, R., "La philosophie biologique d'A. Comte." *Archives d'anthropologie criminelle et de médecine légale,* Oct.–Nov.–Dec. 1909.

87. ——, "La méthode d'étude des affections du langage d'après Hughlings Jackson," *Journal de psychologie* 18 (1921), 752–764.

88. Nélaton, A., *Eléments de pathologie chirurgicale,* 2 vols. Paris, Germer-Baillière, 1847–48.

89. Neuville, H., "Problèmes de races, problèmes vivants"; "Les phénomènes biologiques et la race"; "Caractères somatiques, leur répartition dans l'humanité"; *Encyclopédie française* 7, 1936.

90. Nolf, P., *Notions de physiologie humaine.* 4th ed. Paris, Masson, 1942.

91. Ombredane, A., "Les usages du langage." In *Mélanges Pierre Janet.* Paris, d'Artrey, 1939.

92. Pales, L., "État actuel de la paléopathologie. Contribution à l'étude de la pathologie comparative." Thesis in medicine, Bordeaux, 1929.

92bis. —— and Monglond, "Le taux de la glycémie chez les Noirs en A. E. F. et ses variations avec les états pathologiques," *La Presse médicale* (13 May 1934).

93. Pasteur, L., "Claude Bernard. Idée de l'importance de ses travaux, de son enseignement et de sa méthode," *Le Moniteur universel,* Nov. 1866.

94. Porak, R., *Introduction à l'étude du début des maladies.* Paris, Doin, 1935.

95. Prus, V., *De l'irritation et de la phlegmasie, ou nouvelle doctrine médicale.* Paris, Panckoucke, 1825.

96. Quetelet, A., *Anthropométrie ou mesure des différentes facultés de l'homme.* Brussels, Muquardt, 1871.

97. Rabaud, E., "La tératologie." *Traité de physiologie normale et pathologique,* Vol. XI. Paris, Masson, 1927.

98. Rathery, F., *Quelques idées premières (ou soi-disant telles) sur les maladies de la nutrition.* Paris, Masson, 1940.

99. Renan, E., *L'avenir de la science, Pensées de 1848.* 1890. New ed. Paris, Calmann-Lévy, 1923. (*The Future of Science.* Boston, Roberts Brothers, 1891.)

100. Ribot, Th., "Psychologie," *De la méthode dans les sciences, I,* by Bouasse, Delbet, etc. Paris, Alcan, 1909.

101. Roederer, C., "Le procès de la sacralisation," *Bulletins et mémoires de la Société de Médecine de Paris,* 12 March 1936.

102. Rostand, J., *Claude Bernard. Morceaux choisis.* Paris, Gallimard, 1938.

103. ——, *Hommes de Vérité: Pasteur, Cl. Bernard, Fontenelle, La Rochefoucauld.* Paris, Stock, 1942.

104. Schwartz, A., "L'anaphylaxie." In *Conférences de physiologie médicale sur des sujets d'actualité.* Paris, Masson, 1935.

105. ——, "Le sommeil et les hypnotiques," *Problèmes physio-pathologiques d'actualité.* Paris, Masson, 1939.

106. Sendrail, M., *L'homme et ses maux.* Toulouse, Privat, 1942; reproduced in the *Revue des deux mondes* (15 Jan. 1943).

107. Sigerist, E., *Einführung in die Medizin.* Leipzig, G. Thieme, 1931. (*Man and Medicine: An Introduction to Medical Knowledge.* Translated by Margaret Galt Boise. New York, Norton, 1932.)

108. Singer, Ch., *The Story of Living Things: A Short Account of the Evolution of the Biological Sciences.* New York, Harper, 1931. Later published as *A History of Biology to About the Year 1900.* 3rd and rev. ed. London, Abelard-Schuman, 1962.

109. Sorre, M., *Les fondements biologiques de la géographie humaine.* Paris, Colin, 1943.

110. Strohl, J., "Albrecht von Haller (1708–1777). Gedenkschrift, 1938." In *XVIe Internat. Physiologen-Kongress*, Zurich.

111. Teissier, G., "Intervention." In *Une controverse sur l'évolution. Revue trimestrielle de l'Encyclopédie française*, No. 3, 2nd trimester 1938.

112. Tournade, A., "Les glandes surrénales." In *Traité de physiologie normale et pathologique*. Vol. 4. 2nd ed. Paris, Masson, 1939.

113. Vallois, R.-J., "Les maladies de l'homme préhistorique," *Revue scientifique* (27 Oct. 1934).

114. Vandel, A., "L'évolution du monde animal et l'avenir de la race humaine," *La science et la vie* (Aug. 1942).

115. Vendryès, P., *Vie et probabilité*. Paris, A. Michel, 1942.

116. Virchow, R., "Opinion sur la valeur du microscope," *Gazette hebdomadaire de médecine et de chirurgie*. Vol. II, 16 Feb. 1855.

117. ——, *Die Cellularpathologie*.... Berlin, A. Hirschwald, 1858. (*Cellular Pathology*.... Translated from the 2nd German ed. by Frank Chance. New York, R.M. DeWitt, 1860.)

118. Weiss, A.-G. and Warter, J., "Du rôle primordial joué par le neuogliome dans l'évolution des blessures des nerfs," *La Presse médicale* (13 March 1943).

119. —— and Klein, M., "Physiopathologie et histologie des neurogliomes d'amputation," *Archives de physique biologique*, Vol. XVII, suppl. no. 62, 1943.

120. Wolff, E., "Les bases de la tératogenèse expérimentale des vertébrés amniotes d'après les résultats de méthods directs." Thesis in science, Strasbourg, 1936.

SECTION II

In addition to the works and articles cited as references in the preceding pages, the list below contains other works which provided food for thought.

1. Abrami, P., "Les troubles fonctionnels en pathologie" (Opening lecture of the Medical Pathology Course, Faculty of Medicine, Paris), *La Presse médicale*, 23 December 1936.
2. Amiel, J.-L., "Les mutations: Notions récentes," *Revue française d'études cliniques et biologiques* 10 (1965), 687–690.
3. Bachelard, G., *La terre et les réveries du repos*. Paris, Corti, 1948.
4. Bacq, Z. M., *Principes de physiopathologie et de thérapeutique générales*, 3rd ed. Paris, Masson, 1963.
5. Balint, M., *The Doctor, His Patient and His Illness*. London, Pitman Medical Publishing, 1957.
6. Bergson, H., *Les deux sources de la morale et de la religion*. 20th ed. Paris, Alcan, 1937. (*The Two Soùrces of Morality and Religion*. Translated by R. Ashley Audra and Cloudesley Brereton. Garden City, New York, Doubleday, 1954.)
7. Bernard, Cl., *Introduction à l'étude de la médecine expérimentale*. 1865. Paris, Delagrave, 1898. (*An Introduction to the Study of Experimental Medicine*. Translated by Henry Copley Greene. New York, Macmillan, 1927; New York, Collier, 1961.)
8. ——, *Principes de médecine expérimentale*. Paris, Presses Universitaires de France, 1947.
9. Bonnefoy, S., "L'intolérance héréditaire au fructose." Thesis in medicine, Lyon, 1961.
10. Bösiger, E., "Tendances actuelles de la génétique des populations." *La Biologie, acquisitions récentes. XXVIe Semaine internationale de synthèse*. Paris, Aubier, 1965.
11. Bounoure, Louis, *L'autonomie de l'être vivant*. Paris, Presses Universitaires de France, 1949.
12. Brisset, Ch. *et al.*, *L'inadaptation, phénomène social. Recherches et débat du C.C.I.F.* Paris, Fayard, 1964.

13. Bugard, P., *L'état de maladie*. Paris, Masson, 1964.
14. Canguilhem, G., *La connaissance de la vie*, 2nd ed. Paris, Vrin, 1965.
15. ——, "Le problème des régulations dans l'organisme et dans la société," *Cahiers de l'Alliance Israélite universelle* 92 (Sept.–Oct. 1955).
16. ——, "La pensée de René Leriche," *Revue philosophique* (July–Sept. 1956).
17. ——, "Pathologie et physiologie de la thyroide au XIXe siècle," *Thalés* 9 (1959).
18. ——, *et al.*, "Du développement à l'évolution au XIXe siècle," *Thalès* 11 (1962).
19. Cannon, W.B., *The Wisdom of the Body*. New York, Norton, 1932.
20. Chesterton, G. K., *What's Wrong With the World*. 5th ed. London, Cassell, 1910.
21. Comte, A., *Cours de philosophie positive*. 1838. Vol. III, 48e Leçon. Paris, Schleicher, 1908.
22. ——, *Système de politique positive*. 1852, Vol. II, Chap. V. Paris, Société Positive, 1929. (*System of Positive Polity*. London, Longman, Green and Co., 1875–77. 4 v., Vol. II translated by F. Harrison.
23. Courtès, F., "La médecine militante et la philosophie critique," *Thalès* 9 (1959).
24. Dagognet, F., "Surréalisme thérapeutique et formation des concepts médicaux." In *Hommage à Gaston Bachélard*. Paris, Presses Universitaires de France, 1957.
25. ——, "La cure d'air: Essai sur l'histoire d'une idée en thérapeutique," *Thalès* 10 (1960).
26. ——, *La raison et les remèdes*. Paris, Presses Universitaires de France, 1964.
27. Decourt, Ph. *Phénomènes de Reilly et syndrome général d'adaptation de Selye. Études et Documents*, Vol. I. Tangier, Hesperis, 1951.
28. Duyckaerts, F., *La notion de normal en psychologie clinique*. Paris, Vrin, 1954.
29. Foucault, M., *La naissance de la clinique*. Paris, Presses Universitaires de France, 1962. (*The Birth of the Clinic*. Translated by A. M. Sheridan Smith. New York, Vintage Books, 1975.)
30. Freund, J., *L'essence du politique*. Paris, Sirey, 1965.
31. Garrod. S.-A., *Inborn Errors of Metabolism*. London, H. Frowde, 1909.

32. Gourevitch, M., "A propos de certaines attitudes du public vis-à-vis de la maladie." Thesis in medicine, Paris, 1963.

33. Grmek, M.-D., "La conception de la santé et de la maladie chez Claude Bernard." In *Mélanges Koyré*, I. Paris, Hermann, 1964.

34. Grote, L. R., "Über den Normbegriff im ärztlichen Denken," *Zeitschrift für Konstitutionslehre*, VIII, 5 (24 June 1922).

35. Guiraud, P. J., *La grammaire*. Paris, Presses Universitaires de France, 1958.

36. Huxley, J., *Soviet Genetics and World Science: Lysenko and the Meaning of Heredity*. London, Chatto and Windus, 1949.

37. Ivy, A. C., "What is Normal or Normality?" *Quarterly Bull. Northwestern University Medical School* 18 (1944).

38. Jarry, J.-J. *et al.*, "La notion de 'Norme' dans les examens de santé," *La Presse médicale* (12 February 1966).

39. Kayser, Ch., *Physiologie du travail et du sport*. Paris, Hermann, 1947.

40. —— , "Le maintien de l'équilibre pondéral," *Acta neurovegetativa* 24, 1–4.

41. Klineberg, O., *Tensions Affecting International Understanding: A Survey of Research*. New York, Social Science Research Council, 1950.

42. Lejeune, J., "Leçon inaugurale du cours de génétique fondamentale," *Semaine des hôpitaux* (8 May 1965).

43. Leroi-Gourhan, A., *Le geste et la parole*. I: *Technique et langage*. II: *La mémoire et les rythmes*. Paris, A. Michel, 1964 and 1965.

44. Lesky, E., *Österreichisches Gesundheitswesen im Zeitalter des aufgeklärten Absolutismus*. Vienna, R.-M. Rohrer, 1959.

45. Lévi-Strauss, C., *Tristes tropiques*. Paris, Plon, 1955. (*Tristes tropiques*. Translated by John and Doreen Weightman. New York, Atheneum, 1974.)

46. Lwoff, A., "Le concept d'information dans la biologie moléculaire." In *Le concept d'information dans la science contemporaine*. Paris, Les Editions de Minuit, 1965.

47. Maily, J., *La normalisation*. Paris, Dunod, 1946.

48. Merleau-Ponty, Maurice, *Structure du comportement*. Paris, Presses Universitaires de France, 1942; later edition, 1967. (*The Structure of Behavior*. Boston, Beacon Press, 1966.)

49. Muller, H. J., *Out of the Night: A Biologist's View of the Future*. New York, Vanguard, 1935.

50. Pagès, R., "Aspects élémentaires de l'intervention psycho-sociologique dans les organisations," *Sociologie du travail* 5 (1963), 1.

51. Péquignot, H., *Initiation à la médecine*. Paris, Masson, 1961.

52. Planques, J. and Grèzes-Rueff, Ch., "Le problème de l'homme normal," *Toulouse médical* 8 (1953), 54.

53. Pradines, Maurice, *Traité de psychologie générale*. Paris, Presses Universitaires de France, 1943; later editions 1946, 1948.

54. Raymond, D., *Traité des maladies qu'il est dangereux de guérir*, 1757. New edition by Giraudy, Paris, 1808.

55. Rolleston, S. H., *L'âge, la vie, la maladie*. Paris, Doin, 1926.

56. Ruyer, R., *La cybernetique et l'origine de l'information*. Paris, Flammarion, 1954.

57. Ryle, J. A., "The Meaning of Normal." In *Concepts of Medicine, A Collection of Essays on Aspects of Medicine*. Oxford, Pergamon Press, 1961.

58. Selye, H., "Le syndrome général d'adaptation et les maladies de l'adaptation." *Annales d'endocrinologie*, Nos. 5 and 6 (1964).

59. ——, *The Physiology and Pathology of Exposure to Stress*. Montreal, Acta, 1950.

60. ——, "D'une révolution en pathologie," *La nouvelle revue française* (1 March 1954).

61. Simondon, G., *L'individu et sa genèse physico-biologique*. Paris, Presses Universitaires de France, 1964.

62. Starobinski, J., "Une théorie soviétique de l'origine nerveuse des maladies," *Critique* 47 (April 1951).

63. ——, "Aux origines de la pensée sociologique," *Les temps modernes* (December 1962).

64. Stoetzel, J., "La maladie, le malade et le médecin: Esquisse d'une analyse psychosociale," *Population* 15, No. 4 (1960).

65. Tarde, G., *Les lois de l'imitation*. Paris, Alcan, 1890. (*The Laws of Imitation*. Translated from the 2nd French ed. by Elsie Clews Parsons. New York, Holt, 1903.)

66. Tubiana, M., "Le goitre, conception moderne," *Revue française d'études cliniques et biologiques* (May 1962).

67. Valabrega, J.-P., *La relation thérapeutique: Malade et médecin*. Paris, Flammarion, 1962.

68. Vandel, A., *L'homme et l'évolution*. 1949, 2nd ed., Paris, Gallimard, 1958.

69. ——, "L'évolutionnisme de Teilhard de Chardin," *Etudes philosophiques* (1965), No. 4.

70. Wiener, N., "The Concept of Homeostasis in Medicine." In *Concepts of Medicine: A Collection of Essays on Aspects of Medicine*. Oxford, Pergamon Press, 1961.

71. ——, "L'homme et la machine." In *Le concept d'information dans la science contemporaine*. Paris, Les Editions de Minuit, 1965.

72. Wolff, Etienne, *Les changements de sexe*. Paris, Gallimard, 1946.

73. ——, *La science des monstres*. Paris, Gallimard, 1948.

Index of Names

Zone Books series design by Bruce Mau
Printed and bound by Maple Press